THE ETHICS OF INVENTION

THE NORTON GLOBAL
ETHICS SERIES

General Editor: Kwame Anthony Appiah

PUBLISHED:

FORTHCOMING AUTHORS:
Martha Minow

THE ETHICS
OF INVENTION

Technology
and the
Human Future

Sheila Jasanoff

W. W. NORTON & COMPANY

Independent Publishers Since 1923

NEW YORK * LONDON

For information about permission to reproduce selections from this book,
write to Permissions, W. W. Norton & Company, Inc.,
500 Fifth Avenue, New York, NY 10110

For information about special discounts for bulk purchases, please contact
W. W. Norton Special Sales at specialsales@wwnorton.com or 800-233-4830

Manufacturing by Quad Graphics Fairfield
Production manager: Julia Druskin

Library of Congress Cataloging-in-Publication Data

Names: Jasanoff, Sheila, author.
Title: The ethics of invention : technology and the human future /
Sheila Jasanoff.
Description: First edition. | New York : W.W. Norton & Company, [2016] |
Series: The Norton global ethics series | Includes bibliographical
references and index.
Identifiers: LCCN 2016014456 | ISBN 9780393078992 (hardcover)
Subjects: LCSH: Inventions—Moral and ethical aspects. | Technology—Moral
and ethical aspects.
Classification: LCC T14 J34 2016 | DDC 174/.96—dc23 LC record available at
https://lccn.loc.gov/2016014456

W. W. Norton & Company, Inc.
500 Fifth Avenue, New York, N.Y. 10110
www.wwnorton.com

W. W. Norton & Company Ltd.
Castle House, 75/76 Wells Street, London W1T 3QT

1 2 3 4 5 6 7 8 9 0

For Nina

CONTENTS

THE ETHICS OF INVENTION

Chapter 1

THE POWER
OF TECHNOLOGY

O ur inventions change the world, and the reinvented world changes us. Human life on Earth today looks radically different from just a century ago, thanks in good part to technologies invented in the intervening years. Once firmly earthbound, with only legs and wheels to carry us on land and ships to cross the waters, we have now taken to flight in droves, with more than eight million passengers criss-crossing continents each day in a few airborne hours. If Richard Branson, founder of Virgin Galactic, achieves his dream of building the world's first commercial "spaceline," ordinary people may soon become astronauts. Communication, too, has broken free from shackles of time and distance. When I left India in the mid-1950s, it took three weeks for letters to go back and forth from Kolkata, where I was born, to Scarsdale, New York, where my family first settled. Mail would not arrive reliably. Stamps would be stolen and packages not delivered. Today, an electronic message sent at night from the eastern United States brings an instant reply from a friend in Europe or Asia whose day is just beginning. Facebook connects more than a billion users worldwide with a single mouse click or two.[1] Last but not least, we have cracked the secrets of living and nonliving matter with the decoding of the human genome and the ability to create and deploy a world of novel human-made materials.

Speed, connectivity, and convenience matter a lot, but for most of Earth's seven billion people the quality of life matters more. Here, too, a century of accelerating technological invention has changed us. Work is safer. Air and water in many parts of the world are measurably cleaner. We live appreciably longer. The World Health Organization (WHO) tells us that global "average life expectancy at birth in 1955 was just 48 years; in 1995 it was 65 years; in 2025 it will reach 73 years."[2] Technological innovations account for the trend; better sanitation, drinkable water, vaccines, antibiotics, and more abundant and wholesome food. People not only live longer but enjoy their lives more, through increased access to travel, recreation, varieties of food, and, above all, improvements in health care. If asked whether we would rather be living in 1916 or in 2016, few today would opt for a hundred years ago, even if the world back then had not been racked by war.

Adding up to what some call a second industrial revolution, the technological advances of the past century have propelled wealthy nations to the status of knowledge societies. We have, or are poised to have, unprecedented amounts of information about people's genetic makeup, social habits, and purchasing behavior, and those data are expected to enable new forms of commerce and collective action. State census bureaus are no longer the only bodies that can compile masses of data. Search engines like Google and Yahoo have also become voracious data gatherers, rivaling governments. Even individuals can use devices like Fitbit or the Apple watch to monitor and record volumes of information about their daily activities. Digital technologies have made it possible to combine previously incommensurable forms of data, creating useful convergences between physical, biological, and digital records. What we know about a person today is no longer just a matter of physical descriptors, such as height, weight, ethnicity, and hair color. Nor can people be located through only a few static markers, such as an address and a phone number. Instead, biometric information has proliferated. Passports, for example, can be linked to information gleaned from fingerprints and iris scans collected from anyone who crosses a national border. Apple incorporated a digital fingerprint sensor into its smartphones in the 2010s to replace numerical passcodes and to offer greater security.

The information explosion, spurred by exponential growth in computing capability, now powers economic and social development. The Internet has put unprecedented informational resources at people's fingertips and functions in this respect as an aid to democracy on many levels: for patients wishing to research new drugs and therapies, for small business owners attempting to reach stable markets, or for citizens seeking to pool knowledge about local problems and pressure the authorities to act. Almost everything that people do in high-tech societies leaves informational traces, and these can be consolidated to form astonishingly accurate pictures of their demographic profiles and even their unexpressed desires. From medical environments to commercial ones, the concept of "big data" has begun to expand people's imaginations about what they can learn and how information can open up new markets or provide better public services. In this era, as many governments now recognize, knowledge itself has become an increasingly precious commodity, needing to be mined, stored, and developed like any rare natural resource. The big data age is a frontier for business opportunities, and youthful tech entrepreneurs are the iconic figures driving the new gold rush.

Today's information and communication technologies offer remarkable scope for anyone who can creatively tap into the newly abundant sources of data. Airbnb and Uber took advantage of unused capacity in private homes and private vehicles to turn willing property owners into hoteliers and taxi drivers. When this sharing economy works, everyone benefits because unused capacity is put to use and unmet needs are met more efficiently at lower cost. Families that could not afford to pay for hotels can enjoy dream vacations together without breaking the bank. Enterprises like Uber and Zipcar can help lower the number of cars on the road, thereby reducing fossil fuel use and greenhouse gas emissions. Many of these developments have opened up new frontiers of hope, even in economically depressed regions of the world. Indeed, technology and optimism fit together like hand in glove because both play upon open and unwritten futures, promising release from present ills.

Technological civilization, however, is not just a bed of roses. Offsetting invention's alluring promises are three hard and thorny problems that will frame the remainder of this book. The first is risk, of potentially catastrophic dimensions. If human beings today face existential risks—threats that could annihilate intelligent life on Earth[3]—then these are due to the very same innovations that have made our lives more easy, enjoyable, and productive. Our appetite for fossil fuels, in particular, has created a warming planet where massively disruptive changes in weather patterns, food supplies, and population movements loom uncomfortably close. The threat of total nuclear war has receded a little since the fall of the Iron Curtain, but devastating localized nuclear conflicts remain well within the zone of possibility. Highly successful efforts to manage infectious diseases have produced unruly strains of antimicrobial-resistant organisms that could multiply and cause pandemics. Britain's first "mad cow" crisis of the 1980s offered a sobering preview of the unexpected ways in which poorly regulated agricultural practices can interact with animal and human biology to spread disease.[4] While health and environmental risks dominate our imagination, innovation also disrupts old ways of working and doing business, creating economic risks for those left behind. Taxi companies' bitter opposition to Uber, especially in Europe, reflects an anxiety recently expressed to me on a late-night cab ride in Boston: that the taxi driver is an endangered species.

The second persistent problem is inequality. The benefits of technology remain unevenly distributed, and invention may even widen some of the gaps. Take life expectancy for example. According to the 2013 United Nations World Mortality Report, average life expectancy at birth in rich countries was over seventy-seven years, but in the least-developed countries it was only sixty years, or seventeen years less.[5] Infant mortality rates dropped dramatically between 1990 and 2015, but by WHO estimates rates in Africa remained almost five times higher than those in Europe.[6] Patterns of resource use show similar discrepancies. World Population Balance, a nongovernmental organization dedicated to eliminating poverty, reports that the average American consumed seventeen times as much energy as the average Indian in 2015.[7] In the age of the Internet and instant communication, the U.S. Census Bureau documents wide variation in broadband access within the United States, with 80 percent having such a connection in Massachusetts versus less than 60 percent in Mississippi.[8] The same technologies can be found from Kansas to Kabul, but people experience them differently depending on where they live, how much they earn, how well they are educated, and what they do for a living.

The third problem concerns the meaning and value of nature and, more specifically, human nature. Technological invention upsets continuity. It changes who we are as well as how we live with other lives on Earth, and on this front, too, change is not always felt as beneficial. For more than a century, writers ranging from the German sociologist Max Weber to the American environmentalist Bill McKibben have bemoaned our loss of capacity to wonder at a denatured world, mechanized and disenchanted by technology and threatened by the unstoppable march of progress. The frontiers of disenchantment have widened. Endless new discoveries, especially in the life sciences and technologies, tempt humanity to play out scripts of self-fashioning and control that could transform nature and human nature into manipulable machines. Today's deep ecologists, committed to defending the intrinsic value of nature, want to turn the clock back on some of our most pervasive inventions, such as cars and chemicals. England's Dark Mountain Project, founded and led by the eco-activist Paul Kingsnorth, mobilized around a nightmare of "ecocide," of industrial humanity "destroying much of life on Earth in order to feed its ever-advancing appetites."[9] This collective of writers and creative artists is committed to promoting "uncivilization," through art and literature that might redirect humanity toward less destructive ends.

A more immediate result of technological advancement, other critics claim, is fragmentation and loss of community, in short, the weakening of the social ties that make human lives meaningful. The Harvard political scientist Robert Putnam deplores the America of "bowling alone."[10] This is an America where, in his view, people stay home watching television instead of getting involved in church or civic activities, an America in which women eager for equality and financial independence have left mothering, school teaching, and other community-centered occupations for higher-paid jobs in law offices and corporate boardrooms. Such claims may seem preposterous to today's twenty-somethings who feel connected to increasingly more varied communities through social media. Yet the MIT psychologist Sherry Turkle describes today's youth in America as "alone together,"[11] absorbed in individual, solitary worlds of smartphones and other communication devices, unable to break free and form meaningful, multidimensional, real-world connections.

Technology, in short, has made huge strides in recent decades, but those developments raise ethical, legal, and social quandaries that call for deeper analysis and wiser response. Most visible perhaps is responsibility for risk. Whose duty is it in today's complex societies to foresee or forestall the negative impacts of technology, and do we possess the necessary tools and instruments for forecasting and preventing harm? Inequality raises an equally urgent set of questions. How are technological developments affecting existing gaps in wealth and power, and what steps can be taken to ensure that innovation will not worsen those disparities? A third group of concerns focuses on eroding morally significant commitments to nature and, above all, human nature. Technological developments threaten to destroy cherished landscapes, biological diversity, indeed, the very concept of a natural way of life. New technologies such as gene modification, artificial intelligence, and robotics have the potential to infringe on human dignity and compromise core values of being human. Cutting across all these worries is the pragmatic question whether institutions designed mainly to regulate the physical and environmental risks of technology are up to the task of reflecting deeply enough on the ethics of invention. To examine the complex relationships between our technologies, our societies, and our institutions, and the implications of those relationships for ethics, rights, and human dignity, is the primary purpose of this book.

FREEDOM AND CONSTRAINT

The word "technology" is as capacious as it is unspecific. It covers an astonishing diversity of tools and instruments, products, processes, materials, and systems. A composite of Greek *techne* (skill) and *logos* (study of), "technology" in its earliest usage, back in the seventeenth century, meant the study of skilled craft. Only in the 1930s did the word begin to refer to objects produced through the application of *techne*.[12] Today, the first images the word conjures up are most likely drawn from the world of electronics: computers, cell phones, tablets, software, anything backed by the chips and circuits that make up the silicon world of high-tech societies. But it is well to recall that technologies also include the arsenals of armies, the throbbing dynamos of the manufacturing industry, the plastic forms of genetically modified organisms, the ingenious gadgets of robotics, the invisible products of nanotechnology, the vehicles and infrastructures of contemporary mobility, the lenses of telescopy and microscopy, the rays and scanners of biomedicine, and the entire universe of complex, artificial materials from which almost everything we touch and use is made.

Caught in the routines of daily life, we hardly notice the countless instruments and invisible networks that control what we see, hear, taste, smell, do, and even know and believe. Yet, along with the capacity to enlarge our minds and extend our physical reach, things as ordinary as traffic lights, let alone more sophisticated devices such as cars, computers, cell phones, and contraceptive pills, also govern our desires and, to some degree, channel our thoughts and actions.

In all of its guises, actual or aspirational, technology functions as an instrument of governance. A central theme of this book is that technology, comprising a huge multiplicity of things, rules us much as laws do. It shapes not only the physical world but also the ethical, legal, and social environments in which we live and act. Technology enables some activities while rendering others difficult or impossible. Like rules of the road, it prescribes what we may do without complication and what we do at peril or high social cost. Statins lower blood cholesterol levels and improve cardiovascular health, but people taking statins must be careful to stay away from grapefruits and grapefruit juice. Mac users buy ease and elegance but cannot get behind the computer's built-in design features as easily as PC users can. Purchasers of electric cars drive more climate-friendly vehicles, but they must reckon with the reality that charging stations are few and far between by comparison with gasoline pumps. Food omnivores in rich countries enjoy an unimagined wealth of fresh produce sourced from around the world, but their eating habits leave larger carbon footprints and tax the environment far more than the diets of poor people or committed locavores.[13]

Modern technological systems rival legal constitutions in their power to order and govern society. Both enable and constrain basic human possibilities, and both establish rights and obligations among major social actors. In contemporary societies, moreover, law and technology are thoroughly intertwined. A red traffic light, for example, is a legal and material hybrid whose regulatory power depends on an enforceable traffic code that equates red with stop. Many of modern technology's brightest promises could not be realized without support from the law, such as laws governing contracts, liability, and intellectual property. Conversely, law relies on technology at many points to ensure that its rules will have force and effect, for example, cameras that capture vehicles speeding or police officers shooting. As yet, however, there is no systematic body of thought,

comparable to centuries of legal and political theory, to articulate the principles by which technologies are empowered to rule
us. Nor are the predictive and regulatory instruments invented
over the past half century necessarily strong enough to control
how technology will define humanity's common future.

HOW TECHNOLOGY RULES US

It would be hard to overstate how pervasively technological
inventions rule our actions and expectations. A mundane example may make the point. Until the summer of 2007, I walked
across the T-intersection near my office in Cambridge, Massachusetts, without the benefit of a stoplight. Cars streamed
steadily along two roads, each a major artery into and out of
Harvard Square. I had to gauge how fast they were coming and
when a break was wide enough to permit safe crossing. Sometimes traffic was annoyingly heavy and I had to wait many
minutes; other times I hardly had to pause before stepping out
into the road, though knowing that an aggressive Massachusetts
driver might appear from nowhere as I was hurrying across. I
made personal decisions about when to stop and when to go,
with only my knowledge of local roads and drivers to guide me.

Today, that intersection is regulated. I have to wait up to
three changes of the stoplights before a walk sign lets me cross,
but then I can count on a full nineteen seconds to make it safely
to the other side. For that brief span, recurring on a predictable
cycle, pedestrians own the crossing. Cars stand still as drivers
watch the seconds ticking down, waiting for red to turn green.
The crossing now feels almost sedate. Jaywalking, a student
birthright in this venerable college town, remains an option for
those in a rush, but what used to be a matter of judgment is now

almost a moral question. Should I wait and abide by the law? Should I cross illegally against the light, possibly slowing down a car, getting hit by a bike, or setting a dangerous precedent for other walkers? Inanimate lights backed by invisible experts and unseen electrical circuits have stepped in to discipline behavior that was once risky, individual, and free.

Those traffic lights are a reminder that technologies incorporate both expert and political judgments that are inaccessible to everyday users. Who decided that the crossing needed a light or that nineteen seconds was the right number to keep pedestrians safe and cars moving fast enough? Did Cambridge city officials consult the public? Did they draw the number out of a hat, did they commission their own experts, or did they outsource the design of traffic signals to a consulting company that specializes in such work? The questions multiply. How did the experts, whoever they were, model traffic behavior at the intersection and determine how to allocate time between cars and people? What data did they use, and how reliable was their information? Did they assume all walkers are equally able-bodied, or did they allow extra time for infirm or disabled persons? Normally, we might never think to ask these questions—not unless accidents at the intersection reveal that the experts had made sad mistakes and someone should be called to account.

That observation brings us to the second major theme of this book: governing technology wisely and democratically requires us to look behind the surfaces of machines, at the judgments and choices that shaped how lines were drawn between what is allowed and what is not. Curiously, social theorists have spent a great deal more energy thinking about how to make good laws than about how to design good technological objects, like traffic lights. That asymmetry is puzzling. In democratic societies, uncontrolled delegation of power is seen as a basic threat to freedom. Both legislation and technological design involve delegation, in the first case to lawmakers, in the second to scientists, engineers, and manufacturers. Yet, historically, we have cared a lot more about handing over power to humans than to technological systems. History matters, of course. Philosophers and social scientists have for centuries worried about abuses of monarchical power. The potentially coercive power of technology is a more recent phenomenon. But if we want to retain our human freedoms, our legal and political sophistication needs to evolve along with our technologies. To reclaim human rights in a world governed by technology, we must understand how power is delegated to technological systems. Only then can the delegations be monitored and supervised so as to satisfy our desire for ordered liberty and informed self-governance.

Just as similarities between law and technology are important, so also are the differences. Law on the whole regulates relations between human beings and between people and social institutions. Technology, too, affects interpersonal relations, as for instance when telephones allow salespeople or lobbyists to penetrate private spaces that would be off-limits to physical intruders. But whereas law's efficacy depends on human action and interpretation, technology functions by dividing agency between mindless inanimate objects and mindful animate beings, with far-reaching consequences for responsibility and control. To continue with the traffic light example, a fatal accident at a regulated intersection raises issues of negligence and liability different from those raised by an accident at a crossing without lights. Running a red light is evidence of wrongdoing in and of itself because we have chosen to attribute legal force to the color red. At an intersection without a light, determining who is at fault would require other kinds of evidence, such as the judgment of fallible human bystanders. When to make, and not to make, such potent regulatory delegations to objects remains a deeply ethical question.

AGAINST CONVENTIONAL WISDOM

More than just arrays of inanimate tools, or even large inter-connected systems that facilitate getting things done, new and emerging technologies redraw the boundaries between self and other and nature and artifice. Technological inventions pene-trate our bodies, minds, and social interactions, altering how we relate to others, both human and nonhuman. These changes are not merely material—better cars, computers, or medicines—but transformative of human identity and relationships. They affect the meaning of existence. Our ability to manipulate bio-logical matter, for example, has reframed how we think about life and death, property and privacy, liberty and autonomy. DuPont's 1930s advertising slogan "Better things for better living—through chemistry" sounds hopelessly naïve in an era when life itself, from human bodies to the living planetary envi-ronment, has become an object of design.

Taking this transformative potential into account, this book rejects three widely held but flawed ideas about the relations between technology and society. These are technological deter-minism, technocracy, and unintended consequences. Singly and together, these ideas underpin much of what people commonly believe about the role of technology in society. Each idea offers useful pointers for thinking about how to govern technology well, but each is limited and ultimately misleading. Most dan-gerously, each represents technology as politically neutral and outside the scope of democratic oversight. In this respect, all three notions assert the inevitability of technological progress and the futility of trying to resist, let alone stop, slow, or redirect it. Challenging those presumptions is an essential step in making technology more governable: put differently, these powerful myths need to be set aside in order to reclaim technology for democracy.

The Determinist Fallacy

The idea of "technological determinism" permeates discussions of technological change, even though the term itself may not be familiar to everyone who shares the conventional wisdom. This is the theory that technology, once invented, possesses an unstoppable momentum, reshaping society to fit its insatiable demands. Technological determinism is a common theme in science fiction, where the machine escapes human control and acquires a will of its own. The murderous computer HAL in Arthur C. Clarke's 1968 novel *2001: A Space Odyssey* captured the popular imagination with just this quality of malign intent.[14] Savvy enough to lip-read human speech, but lacking human compassion, HAL kills most of the astronauts accompanying him in the spaceship when he "hears" their plan to disconnect his programmed, amoral mind.

In 2000, Bill Joy, influential cofounder and former chief scientist of Sun Microsystems, wrote a widely noticed article in *Wired* magazine entitled "Why the Future Doesn't Need Us."[15] He argued that the hugely powerful technologies of the twenty-first century—genetics, nanotechnology, and robotics, or GNR as he called them—should be taken far more seriously than most of us do take them because of their annihilating potential. Joy saw a difference between this new era and former times because of the self-replicating potential of the new GNR technologies and the relative ordinariness of the materials needed to produce them. Small groups of individuals with the right know-how could unleash "knowledge-enabled mass destruction." Joy imagined the ultimate dystopian future:

> I think it is no exaggeration to say we are on the cusp of the further perfection of extreme evil, an evil whose possibility spreads well beyond that which weapons of mass destruction bequeathed to the nation-states, on to a surprising and terrible empowerment of extreme individuals.[16]

At first glance, Joy's vision seems not entirely deterministic in that he leaves room for human action and intention, at least by "extreme individuals." But at a deeper level he seems certain that properties intrinsic in the technologies themselves will render the "perfection of extreme evil" a virtual certainty if we continue heedlessly to develop those technologies. Reading this essay by a gifted computer scientist, it is hard to believe that human beings can maintain control over their machines except through inhuman restraint, by not developing the enticing capabilities that also threaten our destruction. Yet there was nothing natural or preordained about HAL's twisted intelligence in *2001*. The murderous computer was the product of earlier, self-conscious human intentions, ambitions, errors, and miscalculations. To treat instruments like HAL as autonomous, possessing an independent capacity to act or to shape action, belittles the ingenuity that designed them and strips away responsibility from the human creators of these marvelous but ungovernable machines.

How can we assess the risks of technology in advance, and is it really true, as Joy fears, that once we launch on the dizzying adventure of technological innovation there can be no turning back? Is there no middle ground for responsible, ethical, technological progress between unbridled enthusiasm and anachronistic Luddism? Joy's essay does not take us to that place, but we can find other handholds on the smooth wall of progress.

Let us return to traffic lights for a moment. In the United States today, we think of each kind of traffic as having a controlled right to an intersection, either through a principled right-of-way that allows some actors always to go before others—as pedestrians on a zebra stripe—or a sequentially regulated right of access for different types of traffic, as at the newly installed lights near my office. This understanding of rights on roadways was itself a novelty less than a hundred years ago.[17] Under English common law, all street users were considered equal, but pedestrians had to yield to motorists as cars took over the urban landscape. Cars caused congestion that could be cured only by giving motorists the right-of-way everywhere except at intersections. Beginning in Texas in the 1920s, traffic signals quickly replaced the more costly human police force. Their rapid spread throughout the world, as devices that carry the same meaning and enforce roughly the same behavior everywhere, is one of the great success stories of modern technological innovation. Once lights were in place, jaywalking became the marked behavior, risky everywhere and punishable with stiff fines in some American states.

But lights, too, have their limitations. They can be overwhelmed by the sheer volume of traffic, and sometimes they are incapable of preventing accidents. At the turn of the twenty-first century, the Dutch road engineer Hans Monderman came up with what seemed like a radically different solution to the problem of mixed traffic on busy roads. His solution was a concept known as "shared space." Monderman's idea was to simplify, not complicate, the technological infrastructure, in short, to do away with lights, signs, barriers, and all other confusing markers. Instead, he proposed that roads and intersections be crafted to encourage users to look out for their own and others' safety. Lights, rails, stripes, and even raised curbs gave way to simpler, more "village-like" designs that encouraged drivers to be more careful. Instead of using material structures to separate different users of the road, Monderman relied on social instincts of respect and care to calm his intersections.

In a 2004 interview with *Wired* magazine, Monderman commented on a traffic circle in the town of Drachten that once had been a nightmare for nonmotorists:

> I love it! Pedestrians and cyclists used to avoid this place, but now, as you see, the cars look out for the cyclists, the cyclists look out for the pedestrians, and everyone looks out for each other. You can't expect traffic signs and street markings to encourage that sort of behavior. You have to build it into the design of the road.[18]

So successful were Monderman's traffic experiments that versions were introduced in metropolitan business districts ranging from London to Berlin. In effect, he had turned the clock back to a kinder, gentler era when all users enjoyed equal rights to the road. It was a small example of how thoughtful, critical reflection on the assumptions underlying a technological system can have enormously liberating effects on those who use the technology, even users who never stopped to ask whether alternatives to inbuilt structures are even imaginable.

If we so often blind ourselves to the power and intent that created our designed and regulated modes of living, it may be because the very word "technology" tempts us to think that the mechanical productions of human ingenuity are independent of their social and cultural foundations. Leo Marx, the eminent American cultural historian, noted that the shift in the word's meaning, from the study of skill to the products of skill, was highly consequential for later thought. It enabled us to conceive of technology "as an ostensibly discrete entity—one capable of becoming a virtually autonomous, all-encompassing agent of change."[19]

That sense of autonomy, however, is illusory and dangerous. A more thoughtful view holds that technologies, far from being independent of human desire and intention, are subservient to social forces all the way through. As Langdon Winner, a prominent critic of technological determinism, pithily put it in a 1980 essay, "artifacts have politics."[20] Many reasons have been put forward for questioning a strictly deterministic position with regard to technologies created by human societies. On the production side, the technologies we make inevitably grow out of historical and cultural circumstances that condition the kinds of needs that societies face, or think they face. Knowledge of the inner workings of the atom did not have to lead to the making of the atomic bomb. That consequence came about because warring states harnessed physicists' theoretical knowledge to make the most destructive weapons that money could buy. Politics influences the uses and adaptations of technology even after products enter the market. Smartphones and social utilities did not cause the Arab uprisings of 2011, although it was fashionable for a time to call them the Twitter revolutions. Rather, existing networks of protest, including the barbarous Islamic State, found phones, video cameras, and services like Twitter useful in giving voice to discontent that had been simmering for years under the region's lids of authoritarian, sectarian politics. These observations have led analysts to rethink technology as both a site and an object of politics. Human values enter into the design of technology. And, further downstream, human values continue to shape the ways in which technologies are put to use, and sometimes even repudiated as in Monderman's calming traffic circles.

The Myth of Technocracy

The idea of "technocracy" recognizes that technological inventions are managed and controlled by human actors, but presumes that only those with specialist knowledge and skills can rise to the task. Who, after all, could imagine approving a new drug without medical knowledge of its impacts on health, or licensing a nuclear power plant without engineering expertise, or running a central bank without training in finance and economics? The belief that modern life is too complicated to be managed by ordinary people has long roots in Europe, dating back to the ideas of the French aristocrat and early socialist thinker Henri de Saint Simon at the beginning of the nineteenth century.[21] Saint Simonisme, as his philosophy was called, stressed the need for a scientific approach to the management of society and a correspondingly authoritative position for trained experts. In the United States at the turn of the twentieth century, the Progressive Era cradled similar beliefs in the inevitability of progress based on science and technology, and the necessary role of experts as advisers at every level of government. By the end of the Second World War, a new dynamic had come into play. Scientists, nurtured by abundant public funding during the war and often relishing their role in affairs of state, lobbied hard to insert more and better science into public decisions. Advisory bodies and positions proliferated, creating in effect a "fifth branch" of the government, beside the traditional legislative, executive, and judicial branches, supplementing the influential "fourth branch" of expert regulatory agencies.[22]

Faith in technocrats and dependence on their skills, however, walked side by side with skepticism and disenchantment. Harold Laski, an influential British economist and political scientist in the twentieth century's brief interwar period, wrote about the limitations of expertise in a 1931 pamphlet that prefigured later doubts and reflections:

> Too often, also, [expertise] lacks humility; and this breeds in its possessors a failure in proportion which makes them fail to see the obvious which is before their very noses. It has, also, a certain caste spirit about it, so that experts tend to neglect all evidence which does not come from those who belong to their own ranks. Above all, perhaps, and this most urgently where human problems are concerned, the expert fails to see that every judgment he makes not purely factual in nature brings with it a scheme of values which has no special validity about it.[23]

Unlike other intellectuals of his day, including some of his American Progressive Era friends, Laski also presciently cautioned against relying too much on eugenics and intelligence tests. Supreme Court Justice Oliver Wendell Holmes, whose increasingly liberal views on the First Amendment were partly shaped by his correspondence with Laski, notoriously supported eugenic sterilization in the 1927 case of *Buck v. Bell*.[24] The excesses of the Holocaust, in which German biologists lent scientific support to Nazi racial doctrines, proved Laski's skepticism to have been tragically well-founded.

Many more recent examples of technological failure, some discussed in detail in later chapters, bear out Laski's long-ago charge that experts overestimate the degree of certainty behind their positions and blind themselves to knowledge and criticism coming from outside their own closed ranks. Although the National Aeronautics and Space Administration (NASA) conducted a highly public inquiry into the 1986 loss of the space shuttle *Challenger*, it did not discover the problems inside the agency that hindered the detection and communication of early warning signs. Only after the loss of a second shuttle, the *Columbia* in 2003, and then only upon consulting with the sociologist Diane Vaughan,[25] did NASA acknowledge deficiencies in its expert institutional culture. In the award-winning documentary *Inside Job*, the director Charles Ferguson detailed how a small circle of well-placed economic advisers, including the "unquestionably brilliant"[26] Lawrence Summers, advocated for financial deregulation and dismissed early warning calls as "Luddite" in the lead-up to the 2008 financial crisis. These examples offer strong arguments for greater transparency and public oversight in expert decisionmaking.

Intended and Unintended Consequences

A third idea that often runs in parallel with technological determinism and technocracy is that of "unintended consequences." It is well known that technologies fail, but it is less obvious who should be blamed for failures and under what circumstances. Indeed, the more dramatic the failure, the less likely we are to accept that it was imagined, let alone intended, by those who designed the object or system. When a toaster breaks down or a car stalls on a freezing morning, we call someone to repair the malfunction; or, knowing that things, like people, have appointed life spans, we grumble a little and buy a replacement for the one that has outlived its usefulness. If an appliance stops working before the end of its warranty period or its expected functional life, we declare it a lemon and ask the seller for our money back. If the malfunction is serious or has caused injury, we may lodge a consumer complaint or go to court to seek redress. Most mishaps in the shared lives of humans and machines are so commonplace that they hardly cause a ripple in society's basic rhythms. We see them almost as natural events, the predictable end points of an aging process that afflicts nonliving as well as living systems.

If technological mishaps, accidents, and disasters seem unintended, it is because the process of designing technologies is rarely exposed to full public view. Anyone who has toured a steel mill, a chocolate factory, or a meatpacking concern knows that visitors must observe strict rules in exchange for even limited transparency. There are places they cannot look and doors they may not open. Once in a while production errors spill into the open with consequences that can be highly embarrassing for the designers, but these are the exception, not the rule. One such episode unfolded in July 2010, when Apple launched its iPhone 4 with great fanfare, selling three million devices in just three weeks. It turned out that, held in a certain way, the phones lost connectivity and refused to function as telephones. The blogs went wild, mocking the legendary company's loss of face and lampooning a product that had taken the market by storm. The event quickly earned a derisive name: Antennagate. It took a smoothly choreographed performance by Apple's late cofounder and chief visionary, the master salesman Steve Jobs, to silence the critics.

Jobs's public meeting and publicity video stressed one theme over and over: "We're not perfect. Phones aren't perfect."[27] Not perfect, perhaps, but as good as it humanly gets, he argued, and in the case of the iPhone 4 far better than any competing gadget on the market. Jobs backed up his claims with a deluge of data on all of the effort that had gone into testing the iPhone antenna's performance. As one blogger noted, it was a bravura performance designed to convince viewers "that the iPhone 4 wasn't just thrown together by some cavemen in a cubicle."[28] If imperfections remained, that was the unavoidable state of a world in which neither humans nor machines can be totally foolproof and fail-safe. Importantly, however, Jobs accepted responsibility on the company's behalf, emphasizing that Apple's people were working overtime to correct the problem and repeating his mantra that Apple wanted all its users to be happy. As a gesture of appeasement, he offered free iPhone covers to compensate early buyers for any inconvenience they had experienced.

The theme of unintended consequences spins technological failures in an altogether different direction. This story line is not the same as Steve Jobs's confession of human and mechanical fallibility. Jobs insisted that Apple's engineers were thinking as hard as possible about a problem and trying to solve it in advance, even if perfect mastery eluded them. By the same token, they would rethink and rejigger their designs if a rare mistake happened. The language of unintended consequences, by contrast, implies that it is neither possible nor needful to think ahead about the kinds of things that eventually go wrong. The phrase is usually invoked after dire or catastrophic events, such as the discovery of the "ozone hole" caused by a few decades of releasing chlorofluorocarbons into the stratosphere, or major industrial disasters such as the lethal 1984 gas leak in Bhopal, India, that killed thousands, or the near-ruinous collapse of global financial markets in 2008. These are moments that reduce us to helplessness, not knowing quite how to respond, let alone how to mitigate the damage. Claiming that such massive breakdowns were unintended assuages the collective sense of paralysis and guilt. It likens the fatal event to a natural disaster or, in insurers' terms, an act of God. It implicitly absolves known actors of responsibility for what went wrong or for picking up the pieces.

The idea of unintended consequences, however, leaves troubling questions hanging. Does it mean that the designers' original intent was not executed as planned, or that things happened outside the scope of their intent because no one could have known in advance how the technology would be used? The two interpretations carry quite different legal and moral implications. In the first case, there may be technology users who can and should be held responsible—like the negligent engineer Robert Martin Sanchez, whose on-the-job texting caused a deadly train accident in Los Angeles in September 2008, killing him and twenty-four others. It is reasonable to think that the telephone's designers did not intend their invention to be used by someone in a job requiring a high degree of sustained attentiveness. Remedies might include changes in law and regulations to enforce higher standards of care by users as well as manufacturers. In the second case, there may be no identifiable people to blame—for instance, when climate scientists discovered decades after the automobile's mass adoption that emissions from cars burning fossil fuel are a major contributor to climate change. The failures here were of imagination, anticipation, oversight, and perhaps uncurbed appetite, none easily cured through preventive policy or improved technological design.

Unfortunately, the two scenarios, though logically distinct, are often hard to disentangle in practice. Part of the problem lies in the fuzziness of the word "unintended." What, after all, are a technology's *intended* consequences? Can the answer be determined only after a failure, in which case it is not very useful for designers or regulators who need to think proactively about possible mishaps? And is there a built-in bias in the use of the term, in the sense that good consequences are always thought to be intended and only bad outcomes are retrospectively labeled unintended? In that case, the talk of unintended consequences serves mainly to reassure us about the fundamentally progressive and beneficent character of technological change. It expresses the hope that no one would intentionally build the potential for catastrophe into devices that are supposed to serve humankind for beneficent purposes.

A second problem is that the term "unintended" seems to fix intention—at least morally relevant intention—at a specific moment in time, though technology in use is never static. The story of the texting train engineer Sanchez illustrates how technologies acquire complex and changing relationships with society long after they have been released into the world. Phones and trains, texting and commuting, may exist in different conceptual spheres for most of us much of the time—as they clearly did for the managers of the Los Angeles commuter rail system before the fatal accident of 2008. But human beings are ingenious, imaginative, and creative users of the devices that the modern world puts within their reach. Whose responsibility is it to track those changing uses? The tragic event in Los Angeles drew attention to a particular synergy that no one had planned for in advance but that supervisors might have found to be quite widespread if they had troubled to look. If one engineer was guilty of such carelessness, then the chances were that others were also behaving in similar ways. Indeed, Kathryn O'Leary Higgins, who chaired the federal inquiry into the fatal crash, declared the use of cell phones by train crews to be a national problem: "This was one day, one train, one crew. It raises for me the question of what the heck else is going on out there?"[29]

STANDPOINT AND METHOD

Technological innovation is as wide-ranging as human diversity itself, and any attempt to make sense of it analytically requires some framing choices. For me, those choices are driven in part by disciplinary training and in part by personal experiences of the world. Much of the empirical material covered in this book derives from my deep familiarity, as a legal scholar, with controversies at the nexus of science, technology, and law. To these examples, I also bring a secondary training in the field of science and technology studies (STS), a relatively new academic discipline dedicated to understanding how the social practices of science and technology relate to those of other institutions in society, such as law. STS analysis demands, in particular, close attention to the ways in which technical and social uncertainties are resolved, and to what gets lost or simplified in the effort to produce believable accounts of the world. Of particular interest to STS are the decisions that sideline doubt and declare that a technological system will function safely and effectively enough. That perspective has produced a robust literature on expert processes, such as risk assessment, that informs the ethical and moral questions raised in this book.

A second methodological orientation comes from my long immersion in cross-national comparisons of science and technology policy and environmental regulation. Those investigations have revealed that countries fully committed to rational, science-based decisionmaking and vigorous protection of public health and safety nevertheless often impose different degrees and kinds of controls on technological systems.[30] Germany is highly antinuclear but France on the whole is more favorably disposed. Neither has access to oil fields or uranium mines within its borders, and their policy discrepancies cannot be explained wholly on the ground of economic and geopolitical interests. American publics, too, do not like nuclear power, at least when the power plant might be situated in their backyards, although most U.S. experts and entrepreneurs are convinced that nuclear power is a viable, indeed essential, safe alternative to fossil fuels.[31] Such divergences point to the importance of history and political culture in ethical reasoning about technological futures—another perspective that informs this book.

The examples I use throughout the book also reflect personal choices and experiences at multiple levels. Many cases involve either the environment or the life sciences and technologies, because these are among my longstanding areas of research interest and because ethical and legal problems were articulated in these contexts sooner than in the digital arena, where issues such as privacy and surveillance are still in flux. I also draw on cases from South Asia, especially India, that reflect a mix of personal and intellectual concerns. India and the United States offer striking similarities and differences that bear on the themes of risk, inequality, and human dignity. Both nations are committed to technological development, both are democratic societies, and both have nurtured respectable traditions of social mobilization and freedom of speech. India, however, is markedly poorer than the United States and more than once has been the site of major technological mishaps and disasters. Juxtapositions of debates in the two countries highlight the challenges of responsible and ethical innovation in an unequal world.

THE RESPONSIBILITY GAP

It would be foolish at best and dangerously innocent at worst to deny the advantages of the human-made instruments and infrastructures that make up the environments of modernity. Yet, whether we treat technology as a passive backdrop for a society that evolves according to unconstrained human choice or attribute to technology superhuman power to shape our destinies, we risk making conceptual errors that threaten our well-being. Centuries of invention have not only made human lives more pampered, independent, and productive; they have also perpetuated forms of oppression and domination for which classical political and social theory barely has names, let alone principles of good government. Unless we understand better how technologies affect basic forms of social interaction, including structures of hierarchy and inequality, words like "democracy" and "citizenship" lose their meaning as compass points for a free society.

The doctrines of technological determinism, technocracy, and unintended consequences tend to remove values, politics, and responsibility out of discussions about technology. Little of moral consequence is left to debate if machines possess their own logics that push society along inevitable pathways. In that case, technocrats argue, rule by experts is the only viable option, since all we want is to ensure that technologies function well, and engineering design and the assessment of technological risks are much too complicated to be left to ordinary people. Further, given the complexities of all large technological systems, there is no realistic alternative to living with uncertain futures containing unforeseeable threats. Viewed through the lens of unintended consequences, many aspects of technology simply cannot be known or effectively guarded against in advance. How could Henry Ford possibly have foreseen climate change when he developed the Model T? Surely it is better for societies to be inventive and creative, take risks in stride, and learn to do better if and when bad things happen!

These arguments can lead to fatalism and despair about those aspects of modern lives that are threatened or constrained by technology. We have seen, however, that conventional wisdom is faulty. Theories that represent technology as intrinsically apolitical or ungovernable have generated countervailing insights that neither technologies nor technical experts stand outside the scope of politics, moral analysis, or governance. The challenge for modern societies is to develop sufficiently powerful and systematic understandings of technology for us to know where the possibilities lie for meaningful political action and responsible governance. The bargains struck in enhancing human capability do not have to be Faustian, ratified between unequal bargaining partners under conditions of blind ignorance or irreducible uncertainty.

But what are the most promising means to ensure that technology will not slip from human control, and what tools, conceptual or practical, can we deploy to hold our proliferating inanimate creations in check? The remainder of this book takes up these questions by looking at the problems of risk, inequality, and human dignity that must be addressed if societies are to live more responsibly with their technological inventions. Chapter 2 looks at the risks that accompany almost all technological innovation, asking where they originate and how they are governed. How do technological risks arise, who assesses them, by what criteria of evidence and proof, and under what sorts of supervision or control? Chapters 3 and 4 address the theme of inequality and the structural foundations of injustice in technologically advanced societies. Chapter 3 examines some dramatic failures in technological systems and inquires how more ethical systems of governance could be devised in a world where risks and expertise are unevenly distributed across national boundaries and responsibility for prediction and compensation is hard to pin down. Chapter 4 traces the controversies over genetic modifications of plants and animals, uncovering the transnational ethical and political dilemmas that arise when scientists tamper with nature on global scales. The next three chapters look in different ways at the evolving role of individual liberty and autonomy in emerging technological systems. Chapter 5 considers the ethical and moral implications of changes in biomedical sciences and technologies, examining the institutions and processes through which decisions are made about the limits of manipulating human biology. Chapter 6 delves into the rapidly expanding world of information and social media, mapping the challenges to privacy and freedom of thought that have arisen in the early decades of the digital revolution. Chapter 7 turns to the vexed question of intellectual property and the rules that govern the tensions between the ideal of free inquiry and the reality of proprietary knowledge. The two final chapters explore questions of control and governance. Chapter 8 reviews various mechanisms that have been devised for engaging publics to play more active roles in the design and management of their technological futures. Chapter 9 returns to the book's opening and ultimate question: how can we restore democratic control over technological forces that appear too rapid, too unpredictable, or too complex to be subject to classical notions of good government?

Chapter 2

RISK AND RESPONSIBILITY

In its simplest definition, technology is a means to an end—or, in the modern era, the application of expert knowledge to achieve practical goals. This understanding of technology, however, obscures a glaring limitation. It implies that the "ends" of invention are known in advance; ingenuity comes into play mainly to accomplish already determined ends. No doubt this was true in the primal stages of human development, when our ancestors were interested mainly in foraging for food and sheltering or defending themselves against weather and enemies. Animals had to be caught on the run or on the fly, so slingshots and arrows were invented; rivers and lakes had to be crossed, so bridges were built and hollowed-out tree trunks floated. Display cases full of the dusty handiwork of ancient toolmakers can be seen in any museum of prehistory—a seemingly inexhaustible supply of sharpened flints, shaped ax heads, stone mortars and pestles, and crude farming implements.

Modern technologies, however, are rarely so one-dimensional. They defy any simple, one-to-one correspondence between means and ends. One reason is that the ends of technology in society are never static, and technological systems evolve much as biological organisms do, along with the societies in

which they are embedded. The most useful and enduring technological inventions are those that find their way into a wide
variety of uses—the wheel, the gear, electricity, the transistor, the
microchip, the personal identification number (PIN). Successful
technologies adapt as social needs mature and values change.[1]
Even the humble lightbulb diversified in form and function as
new materials came on line and people discovered that com-
pact fluorescent and LED bulbs could light homes and cities at
a fraction of the energy used by older incandescents. Today's
cars look vastly different from Henry Ford's Model T because
they have accommodated to users of many stripes, from racing
car enthusiasts to large families with a penchant for camping
or driving long distances. Technologies that do not adapt often
become quietly obsolete. Landline telephones, for example, were
very well suited to one purpose: receiving and transmitting the
human voice from one fixed point to another. Mobile phones
muscled in on the conventional telephone because they not only
enabled calls but were also portable. In turn, these devices were
pushed aside by smartphones that supplement phone calls with
many other functions, such as texting, accessing the Internet,
and taking commemorative pictures of the owner in dramatic
settings.

Technologies that connect means and ends, moreover,
have become more complex. Few if any drivers understand
the software that now controls many of the features that have
made cars come closer to the safe, clean, self-driving (and
self-parking) ideal. But concealed and inaccessible instru-
mentation can cause new kinds of fatal accidents, as when air
bags inflate too violently and without warning. Cars driven
by software can stop dead on highways without the warning
rumbles and sputters that once told us that mufflers or batteries
were failing. In a textbook case of new-age risks, Volkswagen,

one of the world's most respected car manufacturers, disclosed in the fall of 2015 that it had willfully engineered its software to give misleading readouts from diesel engines during emission tests.[2] These "defeat devices" concealed the fact that VW cars performed markedly worse when driving under nontest conditions. Some eleven million consumers were suddenly left with cars that had lower resale value and did not meet applicable air quality standards into the bargain. Volkswagen had not only, by its own admission, knowingly flouted legal requirements; it had disappointed customers' expectations concerning their futures as owners of VW cars.

A loosening of ties between means and ends has made technological pathways less linear and harder to predict. The phenomenon of ozone depletion offers a sobering example. Chlorofluorocarbons came into widespread use as inert, nontoxic, nonflammable refrigerants in the 1960s and then made their way into many new applications, such as aerosol sprays. Only in 1972 did two chemists at the University of California at Los Angeles, F. Sherwood Rowland and Mario Molina, make the surprising discovery that these chemicals were dangerously depleting the stratospheric ozone layer. Academic and industry chemists initially dismissed their finding, and it took more than fifteen years for the pair's work to be accepted and still longer to be recognized with a Nobel Prize. The international community eventually signed a treaty, the Montreal Protocol, to ensure that human societies would not continue to produce or use these life-threatening chemicals, which once had seemed so safe.

For decades, we bundled much of the uncertainty and concern surrounding technological development under the heading of risk. Regulators did their best to assess risks and control them before harm occurred. In its dictionary definition, "risk" means merely the chance of incurring an injury or a loss, like the chances that a passenger will die when flying in a plane or that a homeowner will lose a home in a fire. Technological risks are those that arise specifically from the use and operation of human-made instruments or systems, as opposed to risks from natural disasters that we do not presume to control, such as earthquakes, storms, or epidemic disease. That distinction itself is hard to sustain in an interdependent world. Climate change, for example, crosses the line between natural and technological. It is a phenomenon in nature that humans have caused in an epoch that some now label the anthropocene[3]—defined as the period during which human activity has been a dominant influence on Earth's environment. With the rise and spread of industrialization, risks arising as by-products of human ingenuity have spread to almost every aspect of life. They are here to stay as part of our built environments, and even naturalized into the dynamics of wind, weather, and ocean currents as in the case of climate change. The ideal of zero risk—that is, of a flawlessly functioning technological environment in which machines and devices do exactly what they are supposed to do and no one gets hurt—remains just that: an unattainable dream.

In 1986, the German sociologist Ulrich Beck coined the term "risk society" (*Risikogesellschaft*) to capture the combination of threat, uncertainty, and absence of control that comes to light when technologies fail.[4] The new risks are intangible, like radiation or ozone loss, and nothing in prior experience teaches us how to respond when such a risk materializes. Beck argued that we have entered a phase of modernity in which societies are characterized as much through exposure to various forms of unmanageable risk as through older social classifications such as race, class, or gender. Published in the year of the Chernobyl nuclear accident, Beck's analysis touched deep chords of anxiety in Germany, where his dense, academic, sociological treatise unexpectedly sold many thousands of copies. Its influence, however, proved more lasting than the disaster's aftermath. The book was read as a wake-up call to modern societies to reflect more deeply on, and take more responsibility for, the implications and consequences of their technological adventuring.

Beck called attention to the "organized irresponsibility" of our governing institutions, none of which can claim to understand or control the full extent of the risks that surround us. Science, Beck suggested, can do relatively little to calm the anxieties of this era because new knowledge tends to open up gaping frontiers of doubt and skepticism. Climate science, still in its infancy when Beck wrote his seminal work, provides a compelling illustration of how big advances in knowledge can create important new uncertainties. We know that the Earth is warming through human activity, and we know, too, that extreme weather events will increase in frequency; but how these perturbations will affect regions, let alone the lives and livelihoods of particular socioeconomic groups, is shrouded in uncertainty.

The rest of this chapter looks at the formal governance mechanisms created over the past century to keep technological risks at bay and at their role in empowering or disempowering citizens. This analysis demonstrates that the methods most commonly used to assess risk are not value-neutral but incorporate distinct orientations toward attainable and desirable human futures. One bias that risk assessment begins with is a tacit presumption in favor of change, that what is new should be embraced unless it entails insupportable harm as judged by the standards of today. Another is that good outcomes are knowable in advance, whereas harms are more speculative and hence can be discounted unless calculable and immediate. The chapter begins with an account of some typical maneuvers by which technological failures are reinterpreted, retrospectively, as unintended consequences, thus perpetuating the view that it was only lack of adequate foresight that precipitated harm. I then turn to the techniques of calculation that make potentially ungovernable futures seem governable, by reducing the complexity and ambiguity of the situations within which risks arise. On the whole, as these examples suggest, expert risk assessment tends to value change more than continuity, short-term safety over persistent, longer-term impacts on environment and quality of life, and economic benefits to developers more than justice to other members of society.

NARRATIVES OF REASSURANCE

Technological risk is the product of humans and nonhumans acting together. Any manufactured object, from the most humble to the most complex, can become dangerous if humans mismanage it through poor design or improper handling. Yet, through actual or willed ignorance, we often hold material artifacts blameless by drawing reassuring distinctions between them and their human operators. Technology can then be represented as useful, enabling, and continually progressing, while its adverse consequences are dismissed as unfortunate human errors.

In the case of some widely distributed technologies—guns, cars, recreational drugs, for example—people seem reluctant to blame the invention even if it inflicts a great deal of damage. Guns enjoy a specially protected role in the United States because of the Second Amendment's guarantee of a right to bear arms, even though that provision has been plausibly interpreted as not meant to apply to personal weapons. Though guns are involved in some thirty thousand deaths each year in

America, surpassing even traffic fatalities, the National Rifle Association (NRA), the powerful U.S. gun lobby, famously maintains that "guns don't kill people, people kill people." Casualties happen, the NRA argues, because an intrinsically liberating device has fallen into bad or careless hands. From this point of view, "gun violence" is a misnomer, an unintended by-product of the nation's otherwise laudable gun culture. Some actors are sick or mad or did not handle their weapons responsibly, and the dead or injured sadly happened to be in their sights when the accident happened. In the NRA's trigger-happy imagination, many of those victims might have lived if they, too, had been armed and ready to defend themselves against other people's pathologies.

Failure to aggregate a technology's harmful effects on individual lives may cause significant risks to go unnoticed for long periods of time. Disposable cigarette lighters exploded and caused thousands of injuries and deaths in the 1980s and 1990s. Settled case by case, with court records sealed, these accidents remained invisible until the victims finally sued and forced the industry to adopt stricter standards. Fifteen or so years later, a faulty ignition switch in GM's Chevrolet Cobalt and other small cars caused dozens of deaths and injuries because the air bags failed to deploy. Lawsuits eventually forced the company to reveal longstanding neglect in recalling the vehicles in spite of internal awareness that something was wrong with the switches.[5] More generally, cars continue to crowd our cities and highways, although motor vehicles are involved in more than thirty thousand traffic deaths per year in the United States, about as many as U.S. forces killed in action during the entire Vietnam War. Military technologies are designed to inflict death on inhuman scales, exceeding credible security needs; yet publics reauthorize their production year after year while the global toll of armed conflict is rarely added up and thus gets erased from popular consciousness.

Inquiries carried out in the wake of highly visible technological failures shed even more light on the moves by which societies reassure themselves that just a little more human attentiveness would have kept them safe. The United States lost two space shuttles, the *Challenger* in 1986 and the *Columbia* in 2003, because of uncorrected design defects, in each case killing the seven crew members on board. Costly, tragic, and damaging to the reputation of NASA, the shuttle disasters illustrate several features of risk analysis that matter profoundly for the ethical and democratic governance of technological systems.

First, predicting harm remains an inexact science even for the most thoroughly studied and carefully tested technologies. Causal chains that lead to failure are infinitely varied and hence never fully knowable in advance. In complex systems, even tiny malfunctions can have disastrous consequences because, as the sociologist Charles Perrow noted, the components are so tightly coupled that, once a destructive chain is set in motion, the system is not resilient enough to forgive the initial error.[6] In the case of the *Challenger*, a rubber O-ring lost elasticity at low temperatures and failed to maintain the tight seal needed to contain hot gases away from the shuttle's solid rocket boosters during launch. The escaping gases enveloped an adjacent external fuel tank. Flames, heat, and explosions caused essential joints to bend and break, and the entire vehicle disintegrated less than two minutes from takeoff while horrified audiences watched the event live on their television screens. The crew compartment, even if it survived the breakup without loss of internal pressure, landed in the ocean with such impact that no one could have survived.

In the case of the *Columbia*, a small piece of foam insulation broke off from the shuttle's external tank and hit the edge of the shuttle's left wing, damaging the protective system that guarded the vessel against the heat of reentry into Earth's atmosphere. Senior NASA managers assumed that no midcourse repairs could be made and did not order visual inspection of the damaged area. The fates of vessel and crew were sealed in effect at the moment of the initial debris impact, though the crew remained unaware of what could happen. Two weeks later, barely sixteen minutes before the shuttle's scheduled return to Earth, hot gases penetrated the damaged wing area exactly as feared, causing the vehicle to break up and all crew members to perish.

Second, and more disturbing, the shuttle disasters showed that early warning signals preceded each event but were either not recognized or not respected for what they were. What one knows depends on one's training, inclinations, and position within a technological system. Painstaking reconstruction of the circumstances leading to each tragedy revealed, for example, that engineers knew and were concerned about the defects that caused the lethal accidents well before each of the failed missions.[7] In both cases, some expert voices had warned that things could go badly wrong. Roger Boisjoly, an engineer working for Morton Thiokol, NASA contractor and manufacturer of the *Challenger*'s booster rockets, had recommended against launching the *Challenger* on an exceptionally cold, late January day in Florida. But his superiors at the company and NASA managers eager to get on with an already much-postponed public event overruled Boisjoly. The yea-sayers were reassured by the fact that even though the *Challenger*'s O-rings had malfunctioned in previous flights, on those occasions a second, backup O-ring had kept the seal from failing completely.

Similarly, bits of foam debris had dented the *Columbia*'s body before its ill-fated final flight, but the damage was not severe enough to prompt more precaution. Instead, potentially lethal malfunctions came to be seen as part of an expectable gap between imagined perfection and practical reality, and risks in each case were dismissed as lying within an acceptable range. Only in hindsight did NASA acknowledge that the organization had developed a culture of complacency (a kind of complacency Steve Jobs insisted would never befall Apple after Antennagate), allowing what should have been serious warnings to be reconceived as routine mishaps. The sociologist Diane Vaughan calls this phenomenon the "normalization of deviance."[8] This is a kind of structural forgetfulness that keeps organizations from maintaining adequate levels of vigilance even when they are operating technologies with a high potential for catastrophe.

A third point that the shuttle disasters threw into stark relief is that, as Beck noted, responsibility for the management of complex technological systems is distributed in ways that limit accountability. Parts are manufactured by firms that do not assemble or operate the final product. Operational responsibility, in turn, is frequently split off from managerial decision-making, as illustrated in the *Challenger* case by the division of knowledge and authority between the technical experts at Morton Thiokol and NASA's politically minded leadership. Often, those closest to the details of how things are working in practice, such as the prescient engineer Boisjoly, have no influence over thinking at the top, where technical expertise is likely to be least up-to-date and most subservient to economic and political contingencies. Tragically, both shuttle disasters demonstrated that the consequences of such dispersed oversight, and of the associated mechanical, human, and institutional errors, tend to fall on innocent victims who had little or no say in the decisionmaking that governed their fates.

The difficulty of pinning down responsibility can be seen as the flip side of the myth of unintended consequences. Harm occurs without apparent intention precisely because in so many situations involving technology no single actor is ever in charge of the entire big picture. The twenty-four commuters who died in Los Angeles along with the unfortunate Robert Martin Sanchez could not have known that they were setting out on a fatal journey because no one could have guessed that a train engineer would think of texting on the job. For the passengers, it was a day like any other in a bustling, western megacity, just a day that happened to end tragically. Shuttle launches were never as ordinary as commuter train rides, and each device was among the most carefully built and tested of any machines produced by human beings. Yet, again, no single act or actor was responsible for their doomed flights, or for the deaths of Christa McAuliffe, the schoolteacher who accompanied the *Challenger* crew, and Ilan Ramon, the Israeli air force colonel and son of a Holocaust survivor who flew with the American astronauts on the *Columbia*. The fragmentation of responsibility seen in these individual cases has grown more pervasive with globalization, as we will see again in the next chapter.

RISK AND FORESIGHT

The shuttle disasters were in some respects "normal accidents" in Charles Perrow's terms, but in other respects they were far from normal. Both represented what risk specialists call low-probability, high-consequence events. These are uncommon and only imperfectly predictable, but when they do occur the impacts can be devastating to organizations, communities, and even nations. Loss of life in the *Challenger* and *Columbia* cases was limited all told to fourteen extraordinarily brave individuals, but the economic and psychological effects were profound—for NASA, for the space program, and for a nation trained to think of manned space flight as the ultimate symbol of its technological prowess and worldly standing.

Thanks to extreme foresight before, during, and after the introduction of nuclear power plants, airplanes, and prescription drugs, their respective failures, accidents, and adverse reactions have largely been corralled into the category of low-probability, high-consequence events—though high-impact malfunctions do still happen. Technological risks, however, come in many gradations of probability and consequence. Car accidents continue to claim large numbers of lives annually despite significant improvements in vehicular safety features. In terms of its aggregate effects on society, therefore, automobile use could reasonably be classified as a high-probability, high-consequence risk, even without factoring in the contribution of fossil fuel emissions to global warming and climate change. By contrast, falls from ladders and bicycles, lawn mower accidents, or scalding caused by defective hot water taps are not usually fatal. They impose relatively low costs on society as a whole, even if they cause serious or deadly injury to a significant number of unfortunate individuals each year. People, moreover, generally climb ladders, mow lawns, and turn on the hot water in their homes knowing what to expect of the tools they are using. Accidents involving these, then, can be seen as examples of high-probability but relatively low-consequence events. They differ from low-probability, high-consequence risks not only in their frequency and socioeconomic impacts but also in the underlying relations of familiarity between the users and the instruments used.

TABLE 1: TYPOLOGY OF RISKS

Probability / Consequence	Low	High
Low	Brief power outages (in developed countries)	Noxious odors Falls from ladders
High	Vaccine side effects Nuclear power plant accidents	Traffic fatalities Gun deaths (U.S.)

The makers and consumers of complex technologies operate on very different playing fields with respect to how much they know of the risks involved and whether they are able to act preventively. People living near a nuclear power plant are generally ignorant of a plant's safety record and unable to protect themselves in case of accidents. Airplane passengers do not know about their pilot's flying experience or mental health, though these could affect their lives, as when in 2015 the Germanwings copilot Andreas Lubitz locked his partner out from the cockpit and drove his plane into the Pyrenees, killing everyone on board. Patients using prescription drugs may have a lot of information but still be unable to predict or prevent severe adverse reactions in themselves or their children. Any ethically acceptable system of risk governance has to take into account these wide variations in the causes, consequences, and knowledge distribution across the spectrum of hazardous technologies.

To cope with these challenges, the governments of technologically advanced nations have taken on a virtually universal commitment to regulatory risk assessment: that is, to systematic, public analysis of risk before citizens are exposed to grave or widespread harm, followed by regulation as needed to reduce those risks. Conceptually, this development owes a great deal to the rise of the insurance industry and its partnership with state welfare programs from the middle of the nineteenth century. The principle of risk spreading—taking small amounts of money from all members of a risk-exposed group in order to compensate the few who actually come to harm—has ancient roots. The Babylonian Code of Hammurabi offered rudimentary insurance for farmers, providing that a debtor's loan would be forgiven in a year when a storm or drought destroyed his crop or caused it to fail. Today, hardly any social activity that could cause harm occurs without someone insuring it. The California-based Fireman's Insurance Fund, for example, insures the film industry against injuries from stunt acting and simulated disasters.

Only with the emergence of the modern nation-state, however, did governments begin routinely to insure their populations against collective problems, such as occupational hazards, automobile accidents, aging and unemployment, and sickness or disability. These were harms that anyone might encounter at some point in the course of a life, but—depending on the afflicted person's position in society—with different degrees of damage to livelihood and social standing. Otto von Bismarck, the founder of the modern German nation-state, famously established a series of pension and insurance schemes that became models for other European welfare states. Even in the United States, despite chronic political opposition to the welfare state model, state after state enacted workers' compensation laws at the turn of the twentieth century to provide automatic coverage for employees injured on the job. The Affordable Care Act of 2010, the landmark legislative accomplishment of the Obama presidency, represents yet another extension of the insurance principle, this time to cover the health of all Americans.

Risk assessment is a policy instrument, not an end in itself; it typically precedes and informs regulatory measures designed to protect health, safety, and the environment. Like social insurance schemes, regulatory risk assessment builds on the principle that the probability of harm to affected groups can be calculated in advance. Here, however, the object is not to generate enough money to pay individuals for injury and loss (the primary goal for both public and private insurance) but to make sure that the aggregate amount of harm from major technological developments remains within bounds that society finds tolerable. In conducting risk assessment for hazardous technologies, states affirm a very general responsibility to protect the lives and health of citizens against human-made hazards. Under U.S. laws, drugs, pesticides, food additives, and medical devices may not be marketed without a prior assessment of the risks they pose, and then in most cases only with explicit governmental approval.

By assuming the burdens of assessing and mitigating these risks, state-sponsored risk assessment counters at least in principle Ulrich Beck's charge of organized irresponsibility. It also serves a democratic function. In many nations, administrative procedures require public disclosure of information about technological risks and give interested parties an opportunity to question state and corporate risk experts. Regulatory risk assessment thus provides a window into the inner workings of technology that is not otherwise available in the course of industrial production. It confers on citizens a right to know about some of the invisible risks of modernity.

How does the apparatus of technological risk assessment work in practice? Much has happened since the middle of the twentieth century to routinize the process and make it seem familiar and feasible. In 1983, responding to protracted political controversies over risk-based regulatory decisions,[9] the U.S. National Research Council (NRC) issued a report systematizing the process of risk assessment for all federal agencies entrusted with evaluating and regulating risk. Admirably clear and concise, the NRC's analytic framework was widely taken up both nationally and internationally.[10] At its core was the injunction that "risk assessment" be treated as a largely scientific exercise, insulated from the subsequent process of making value-based decisions, which the NRC termed "risk management." Risk assessment according to this scheme was to take place in discrete, linear steps, each based only on available scientific evidence and all carried out before the development of policy options. The key steps were hazard identification, dose-response assessment, exposure assessment, and risk characterization.

On its face, the NRC proposal made a great deal of sense. If understanding risk is primarily a matter of learning and evaluating facts about nature and society, then it is important to shield that fact-finding process against political manipulation. Hence it seemed like sensible policy to keep risk assessment separate from risk management. The four stages of risk assessment outlined in the 1983 NRC report seemed equally straightforward. Recognition that a hazard exists is the logical starting point for any attempt to minimize or eliminate risk. In a period of rising national concern with cancer, the NRC authors took the harmful health effects of environmental chemicals as their paradigmatic example. Not surprisingly, therefore, the second step in the risk assessment process drew its inspiration from the science of toxicology, in which harm is always seen as proportional to dose. Dose-response assessment is the scientific attempt to determine how bodily exposure relates to the severity of an impact, a calculation that applies to radiation as well as to toxic chemicals. Exposure assessment, the third step in the NRC framework, measures a hazard's impact on a given population by studying who is exposed and by what pathways. Ubiquitous compounds such as air pollutants and persistent substances such as lead entail higher levels of population exposure than, for example, specialty drugs used to treat small and limited classes of patients. Finally, the step of risk characterization estimates the amount and distribution of harm at varying levels of exposure throughout a population. This may include the measurement of harm to specially vulnerable groups, such as children, pregnant mothers, or people with preexisting health conditions. This determination in turn helps define reasonable policy options for reducing overall levels of risk to society.

FRAMING RISKS: A QUESTION OF VALUES

Rational as it seems, the NRC's risk assessment framework has drawn well-justified criticism for the subtle and not so subtle ways in which risk assessment narrows the questions that get asked about technological innovation. The most basic limitation is that risk assessment frames problems in ways that may not produce optimal results for collective well-being. The American sociologist Erving Goffman popularized the concept of framing in sociology and social theory.[11] Framing is a widely recognized social process through which individuals and groups make sense of previously unorganized experience. Goffman pointed out that even the most commonplace happenings mean nothing to us unless we can identify and label their elements and relate them to one another. A Martian or a person from a pre-automotive culture, for example, may not know the vernacular meaning of running a red light; you need to be a member of the global driving culture to interpret that action as unlawful.

But in the effort to make sense, framing may accentuate biases that are already built into a social system, and this can cause significant ethical problems.

Framing inevitably calls for a sorting of sensory experiences into different layers of significance: some elements belong within a story, others do not, and the totality makes sense only if the elements are arranged properly. We go out for an evening walk and pass a stranger. She seems to be talking to herself, a sign of mental disturbance we believe, and we cross the street in avoidance. We go out for an evening walk and pass the same stranger, again talking to herself, but this time we are primed to look for the earphone showing that she is talking to somebody other than herself. Suddenly there is nothing to fear. The situation now seems normal because it has been fitted into a pattern of conduct that arose with the cell phone and that rendered previously odd behavior (talking to no visible partner in public) understandable and unthreatening (talking to an invisible but real partner in public).

What we see as normal and what we seek to explain as abnormal are neither objective nor neutral perceptions. They reflect deep-seated cultural predispositions. The fatal shooting of the unarmed African American teenager Trayvon Martin in February 2012 in a gated Florida community illustrates the power of culture in framing. As many commentators pointed out, the Hispanic neighborhood watcher George Zimmerman might not have identified, followed, or got into a deadly scuffle with a white youth in the same setting. Race in this context created a frame of suspicion, with tragic consequences. Who frames a risky situation and to what end are thus political questions through and through, and politics can shift the frame for subsequent similar events. The later shooting deaths of the unarmed teenagers Renisha McBride in a Michigan suburb in 2013 and Michael Brown in Ferguson, Missouri, in 2014 sparked a nationwide debate and in Brown's case provoked sustained civil unrest in his hometown. McBride's killer, a civilian, received a seventeen-year prison sentence; Brown's killer, a policeman who had been punched by Brown, was not indicted. These cases displayed a frame shift in motion—a society less willing to take losses of young black lives during law enforcement in stride, to the point of launching the Black Lives Matter movement, but at the same time not prepared to penalize police officers unduly for the use of deadly force.

A second drawback of conventional risk assessment is that it ignores risk's systemic and distributive consequences. Challenging Beck's universalizing notion of the risk society, hazardous technologies do not in practice spread randomly or evenly across space and social groups. They tend to cluster in poorer, politically disadvantaged neighborhoods (and, increasingly, in poorer regions of the world), where people have neither the technological expertise nor the connections and clout needed to keep dangerous developments out of their localities. During the 1970s, many better-off communities throughout the industrialized world organized to exclude hazardous or otherwise noxious technological facilities, such as incinerators, airports, and nuclear power plants. These efforts were widespread enough to acquire a name of their own: the NIMBY ("not in my backyard") syndrome. Citizens of Ithaca, New York, for example, mobilized scientists from Cornell University to stop the New York State Gas and Electric Company from building a nuclear power plant on picturesque Cayuga Lake.[12] Cambridge, Massachusetts, established a scientific review committee to advise on the appropriateness of conducting high-risk genetic engineering research in their city. The committee determined that the most hazardous experiments (at the so-called P4 or strictest level of containment) should not be permitted within city limits.[13] Both cases illustrate the success of the NIMBY syndrome when backed by the expertise that citizens can muster in university towns, but on a national or even regional scale such resources are less accessible. Accordingly, enterprises rejected by those with the capacity to mount good technical arguments and enjoying good access to political power tend to move to places where those resources are lacking.

A third criticism centers on the ways in which conventional risk assessment restricts the range of evidence that enters into the analytic process. Almost by definition, regulators conducting risk assessment are forced to ignore knowledge that does not look like science as it is usually understood, that is, knowledge gained through publication in peer-reviewed journals or produced through authorized expert advisory processes.[14] This means that the knowledge of ordinary citizens, which may be based on long historical experience and repeatedly verified by communal observation, tends to be set aside as subjective or biased, and hence as mere belief rather than reliable evidence.[15] Such experiential knowledge, however, can be especially valuable when it is based on direct interactions with machines or natural environments: industrial workers may understand the risks of their workplace better than the design engineers, and farmers know the cycles of crop behavior in their fields better than global climate modelers. Because of its focus on specific, identifiable hazards, moreover, the risk assessment framework tends to emphasize single causes, such as individual chemicals or emissions from particular facilities, thus underestimating the possible synergistic effects of multiple harmful exposures. Yet in daily life a person tends to encounter risk in many different forms and from more than one source. Those combined exposures may have effects that remain completely outside the risk assessor's frame of analysis.

Complexity is difficult to capture through the typical linear model of risk assessment, which begins with a single, well-defined hazard and traces its path through population-wide exposure studies. We know that life spans are shorter and health indicators worse in lower-income neighborhoods, where people are exposed to a barrage of physical and psychological insults, from poor nutrition and drugs and alcohol abuse to the stresses of joblessness, broken marriages, mental illness, and urban violence. The neatly compartmentalized framework of risk assessment, with its emphasis on peer-reviewed scientific studies, focusing on one exposure at a time, provides little or no purchase point on the complex social and natural causes that create zones of concentrated misery. Ignorance exacerbates the inequalities. The places where risks tend to cluster are also the places least endowed with resources to generate reliable data about baseline physical, biological, and social conditions.

The fourth, and in some ways most troubling, problem with the NRC's approach to governing risks is that it leaves little or no space for public deliberation on benefits while a technology is still on the drawing boards or in its infancy, before substantial economic and material investments have made retreat or substantial redesign unthinkable. Benefit calculations, usually in the form of cost-benefit analysis, enter into the picture relatively far downstream in the NRC model, only during the stage of risk management. Put differently, the risk assessment–risk management paradigm does not allow for early public consideration of alternatives to a designated technological pathway, including choices not to proceed or to proceed only with caution on a slower timetable.

Risk assessors, for example, are not typically in a position to compare risks and benefits across alternative scenarios. In the case of nuclear waste management, they are allowed to question whether high-level radioactive wastes can be safely stored in a permanent geological repository at a particular location, such as Nevada's Yucca Mountain. They are not authorized to undertake full-blown studies of alternatives, such as leaving the wastes where they are, in more than a hundred temporary, relatively insecure facilities throughout the country, or shipping them to another nation that is more inclined to accept them. U.S. nuclear risk assessors never confronted the extreme scenario of altogether phasing out the nuclear option, as was done with varying results in Austria and Germany. Whether to say no to a technological pathway is seen in the United States as a political, not a scientific, question and hence is not a part of standard risk assessment.

Yet experience suggests that attempts to segregate underlying political values away from technical analyses of risk are often counterproductive and create expensive, intractable controversies. The United States spent more than two decades and ten billion dollars studying the Yucca Mountain repository before the Obama administration declared its intention to abandon the project in 2009. During that contorted history, Nevadans never acquiesced in having their state used as the nation's nuclear waste dump. The decision to drop Yucca Mountain reflected the U.S. Department of Energy's tacit conclusion that scientific analysis alone could not shut down the intractable ethical and political conundrums of creating a single nuclear waste repository, with one state bearing the brunt of an entire nation's appetite for energy. Challenges to factual claims in this context are merely the presenting symptom of unresolved polarization in social preferences. In such contested territory, science alone could never produce watertight, publicly acceptable demonstrations of safety. More is needed to close off debate, a coming together on purposes, meanings, and values.

A POLITICAL CALCULUS

Second only to framing, quantification is key to the utility of risk assessment. By systematically attaching mathematical probabilities to different types of harm, risk assessors are able to compare threats from extremely diverse sources: air pollution versus unemployment; loss of biodiversity versus the risk of famine; prevention of malaria versus the decimation of songbird populations. In turn, these risk-to-risk comparisons give policymakers a seemingly robust basis for deciding which risks are most worth controlling and, in a universe of necessarily limited resources, how stringently each should be controlled. These virtues help explain why risk assessment became in the last quarter of the twentieth century one of the most popular instruments for controlling the hazards of technology, but setting numerical values on the probability of harm adds another layer of complexity to the politics of risk assessment.

Assigning numbers to a scenario is itself a means of framing and therefore is also a political act. Almost by definition, quantification downplays those dimensions of an issue that are not easily quantifiable, even though they may be crucially important for public welfare. Senator Robert F. Kennedy spoke eloquently about this aspect of enumeration when he criticized the gross national product (GNP) during his 1968 campaign for the presidency:

> Yet the gross national product does not allow for the health of our children, the quality of their education, or the joy of their

play. It does not include the beauty of our poetry or the strength
of our marriages; the intelligence of our public debate or the
integrity of our public officials. It measures neither our wit nor
our courage; neither our wisdom nor our learning; neither our
compassion nor our devotion to our country; it measures everything, in short, except that which makes life worthwhile. And it
tells us everything about America except why we are proud that
we are Americans.

Similar statements can be made about the quantification of technological risks and benefits on lesser scales than the national
GDP, with the further proviso that choices of what to measure
or not to measure can have immediate and direct impacts on
people's lives and health.

The methods used to evaluate many chemical and biological products—pesticides, food additives, pharmaceutical drugs,
chemicals in the workplace, and genetically modified organisms (GMOs)—shed light on some of the limitations of using
numbers as the primary means for forecasting futures. Quantitative risk assessment measures very particular kinds of harms,
usually a known disease or injury from exposure to a specific
hazard source. Thus, risk assessment methods can estimate
with a high degree of accuracy the probability of increased cancer causation in a person exposed at varying concentrations to
a well-characterized toxic chemical, such as benzene or vinyl
chloride. Risk assessment can produce plausible estimates of
the likelihood that potentially damaging genes will be transferred from a GMO to a related, nonmodified species or to
organisms that were never targeted as recipients of that gene.
Such assessments are routinely used as a basis for regulating
hazards: for example, setting standards for permissible concentrations of benzene or vinyl chloride in the workplace or

the ambient air; or determining the distances to be maintained between farmed areas planted with GMOs and areas that must be kept GMO-free.

Such typical risk assessment exercises, however, invariably leave out some issues of concern to people. In the case of chemicals, for example, we learned through hard experience that product-by-product risk assessments do not attend to the synergistic effects of multiple pollutants, especially when these are concentrated in poor or powerless communities that cannot guard themselves against an influx of hazardous industries. This recognition led U.S. President Bill Clinton in 1994 to issue an executive order (a rule applying to operations of the executive branch of government) requiring each federal agency to "make achieving environmental justice part of its mission." Under that mandate, U.S. agencies must identify and address the "disproportionately high and adverse human health or environmental effects" of their programs and policies on minority and low-income populations. The point here is that environmental justice was a relative latecomer to the field of risk management policy although it makes sense to treat this principle as fundamental in a world where each and every human depends on the environment for well-being and survival. It was not initially foreseen that quantitative risk assessments aiming to minimize risk for "standard" bodies at "standard" exposures would not assure the same levels of protection for all people in all communities. Policy approaches designed primarily to ensure safety for individuals in effect masked the deeper structural politics of risk inequality.

Other important issues, too, have been regularly sidelined in policymaking based on quantitative risk assessment. One is the matter of relative costs and benefits. Assessments are typically undertaken only when a technological project, such as developing or marketing a new product, is already well under-way. Risk assessors have no obligation, and indeed no factual basis, to consider whether lesser-impact alternatives might be available. Accordingly, alternatives including the null option of doing without a product or process altogether are never directly under investigation. Nor are the relative risks and benefits of different ways of achieving the same results usually compared. New chemical pesticides are thus approved one at a time, pro-vided they do not pose unreasonable risks to human health or the environment; risk assessment does not set pesticide use itself side by side with the benefits and risks of using radically different approaches to pest management. GMOs, similarly, are assessed in and of themselves but not (as we will see again in chapter 5) in relation to alternative forms of agricultural production.*

A still more basic omission from most quantitative risk assess-ments is the social behavior that makes technology work. We know, of course, that the potential for harm lodges not solely in the inanimate components of technological systems but in the myriad ways that people interact with objects. Cigarettes are not harmful till people smoke them. Guns do not kill unless people fire them. Cars must be driven to cause crashes or release carbon into the air. Chemicals need to be taken off the shelf and used in order to get into the air, the water, or the food chain. The risks of modernity, in other words, are not purely technological—mere malfunctions within machines—but hybrid, dynamic, and sociotechnical. Yet risk assessment only

---

*The biotechnology industry has long claimed that GM crops, with geneti-cally engineered pest resistance, are more natural and less environmentally damaging than chemical pesticides. But this attempt to carve out a market niche for a different kind of industrial agriculture has not been tested along-side various competing alternatives, such as organic food production or inte-grated pest management.

imperfectly takes account of technology's embeddedness in the workings of society.⌋

Some elementary aspects of human behavior are factored into the exposure assessment stage of risk analysis. In calculating the duration, intensity, and extent of individual and population contact with a given hazard, the risk assessor has to consider people's likely conduct. For instance, if the at-risk group consists of neighbors of a polluting plant, will they live at the same distance from the plant for an entire lifetime or will they more likely move away after some years? Numerical answers to such questions can be given in principle, but there is much room here for oversimplification and inaccuracy. As in the garment industry, the "typical" person whose exposure we want to measure is a useful fiction. One of my own small and relatively harmless encounters with the irrationality of machines tells the story. When I tried to establish my trusted traveler status with the U.S. border security authorities, the government's fingerprint reader refused to record my fingertips properly on the first two tries. Eventually, a helpful officer explained to me that "they sometimes don't work so well for females." (The cure, it turned out, was to use hand lotion before pressing my fingers to the screen.) Not all machine designs are so innocently stupid. Tragedy may strike if safety features intended for a standard user encounter a nonstandard body, as happened in the United States when air bags meant to protect adult front-seat passengers inflated too forcefully and ended up killing dozens of small children.

More complicated questions about the interactivity between human and technological systems tend to be completely glossed over in risk analysis. How, for instance, should one factor into risk assessment that people use technologies with very different levels of physical skill and intellectual maturity? In the designing of safe automobiles, should the imagined driver be a staid, middle-aged parent, a reckless teenager, a habitual drinker, a race car enthusiast, an elderly person with diminished vision, or an artificial hybrid of all these? Given that so much of technology today crosses regions and political boundaries, how can the manufacturer's risk assessment properly accommodate the potentially very different knowledge and habits of end users in distant countries? The simplest omissions can have lethal consequences. Pesticide warning labels written in English, for instance, have failed to warn farmers in non-English-speaking countries. In reverse, when assessing the risks of imported technologies, can numbers capture the accuracy and reliability, or the honesty, of the suppliers of the technology or of the information needed to assess it properly? The notorious Volkswagen "defeat device" case (see chapter 2) certainly suggests otherwise.

In sum, risk analysis as it came to be practiced by policymakers in the last third of the twentieth century is a major source of trouble for democracy. It rests on relatively thin descriptions of reality. It tends to arrive too late in the process of technological development and operates with a shallow awareness of the complexities of sociotechnical systems. It constructs typical or standard scenarios that reduce the complexity and dynamism of real-world sociotechnical interactions. It is ahistorical and unreflective about its own normative assumptions. To improve the governance of technology, we need to take better account of the full range of values that humans care about when contemplating the future—not just the value of change but also that of continuity, not just physical safety but also the quality of life, and not just economic benefits but also social justice. Let us turn in the following chapters to some examples that illustrate the limits and failures of calculative rationality in governing particular sectors of technology.

Chapter 3

THE ETHICAL ANATOMY
OF DISASTERS

Disasters are dramatic examples of humanity's failure to live in harmony with the products of its technological ingenuity. Often coupled to place-names, disasters echo through history as sites of calamity: the poisonous dioxin release in Seveso, near Milan in northern Italy, that inflamed children's skin and killed thousands of animals in 1976; the deadly blanket of methyl isocyanate that enveloped the sprawling city of Bhopal in central India in 1984; the catastrophic nuclear accidents in the Ukrainian city of Chernobyl in 1986 and Japan's Fukushima in 2011; and their less damaging but still foreshadowing precursors at Britain's Windscale, later renamed Sellafield, in 1957, and at the Three Mile Island power plant near Harrisburg, Pennsylvania, in 1979. A list of lesser-known place-names can be recited in any industrial nation anywhere in the world, commemorating local tragedies at coal mines, oil rigs, highways, railway crossings, stadiums, dams, bridges, flyovers, and factories. The count of lives lost, environments poisoned, businesses ruined, and bodies made ill or useless by technological accidents will never be complete.

Technological disasters stand for more than failure and loss. They are also morality tales about carelessness and overreaching. Bound up in the story lines of these events are lessons about the kinds of mistakes we most often make in our dealings with technology and the sorts of consequences we are most likely not to foresee. Such lessons are particularly important in an age when, partly as a result of increasing wealth and population density and partly because of the mobility of capital and industry, the likely impact of disasters has risen while their causes have become less easy to anticipate and pin down. Disasters, too, offer insights into the dynamics of social inequality. As such, they are prime subject matter for ethical and political analysis.

Statistics tell one common, troubling story about disasters: the poor suffer more than the rich. The poor work longer in more dangerous conditions, live in less protected dwellings, and have fewer resources with which to defend themselves when disaster strikes. Often, their communities are more marginal, unable to pressure officials to guard against loss and suffering. An accident in a textile factory in Bangladesh in the spring of 2013 tragically played out this familiar script. On April 24, the nine-story Rana Plaza building in an industrial suburb of Dhaka collapsed, killing a reported 1,129 workers within.[1] The surprise was not that the building fell but that it was being used as a workplace for thousands in spite of every possible warning. The Rana Plaza's imposing blue glass entrance façade and wide front steps presented a misleading front to the world. It was built of substandard materials, too much sand mixed in with the concrete, on swampy soil, with questionable clearances from inspectors and political authorities.[2] An illegal eighth floor had been added atop an already fragile structure and a ninth was under construction when the building fell. Heavy generators operating on upper floors provided electrical backup during Dhaka's frequent power outages. When on, these machines shook the building with their vibrations, weakening walls and supports, fatally as it happened on that April morning.

The day before the disaster, all the workers were evacuated when an engineer discovered alarming cracks in the building's walls and load-bearing columns. Factory owners and some 3,500 garment workers waited anxiously outside on the morning of the accident to learn whether they could go in. Quotas had to be met and contracts satisfied. Two people, one a structural engineer, were called in to inspect the cracks, and they declared the building safe. Some later said they were in the pay of Sohel Rana, the building's owner, a thirty-five-year-old, gangland-style boss, whose association with the Awami League, Bangladesh's majority political party, helped legitimate his shady dealings in real estate and, reportedly, drugs. At 8:30 a.m., half an hour after the workers trooped in on time, the power went down and the generators began their deadly work. In barely ten minutes, one corner of the structure collapsed, followed minutes later by the entire building. Rana himself was in the building's basement at the time. He survived and attempted to flee from justice but was caught near the Indian border and returned to Dhaka to stand trial for murder.

Until Sohel Rana's capture and imprisonment, the loss of the building bearing his name seemed to confirm every sad expectation about risk and poverty. The poor labor for small gains, working under conditions of Dickensian injustice and misery, while the rich profit and get richer. Production and consumption are separated by long supply chains, as Bangladeshi hands, working for wages as low as fifteen dollars a month, sew zippers and shirt collars for purchasers earning many hundreds of times as much in distant Europe and America. Such discrepancies in wealth and power between points of origin and points of sale open up spaces of corruption in which payoffs such as Rana's bribes to political officials are routine. When things go desperately wrong, it is the poor who bear the brunt, sometimes dying in large numbers, while the rich live to see another day.

In the technological world, however, risk is not inevitably kind to the rich. We are all to some extent vulnerable denizens of risk societies. A lethal outbreak of food-borne *E. coli* poisoning that occurred in Germany in 2011 offers an instructive counterpoint to the Rana Plaza story. Beginning on May 21 of that year, German health authorities began reporting increased cases of acute diarrhea, especially among young women, and a condition known as hemolytic-uremic syndrome that often follows such infections and may lead to kidney failure and death.[3] By the time officials declared the outbreak over, in late July, more than four thousand cases of the two diseases had been reported throughout Europe, the great majority of them concentrated around the city of Hamburg in northern Germany. By the time the crisis ended, fifty-three people had died, all but two in Germany.[4] Many living victims were left permanently impaired and in need of long-term health care such as dialysis. It was the worst documented case of *E. coli* infection in two generations, and the economic costs, estimated at $1.3 billion for farmers and industries alone by the World Health Organization, most likely topped those of any previous instances of food-borne illness in history.

How could such a devastating public health crisis occur in one of the world's richest, most health conscious, and most technologically sophisticated countries? Early difficulties in identifying the causes of the infection underline the dispersed character of food production and consumers' as well as regulators' lack of control over the supply chain that brings food to tables, even in the wealthy West. Initial efforts to pin down causes, on the basis of comparisons between affected and unaffected parties eating in the same locations, suggested a link to fresh salads, a finding that squared with the disproportionately high number of adult women affected. This led to an official advisory not to eat uncooked salad ingredients, especially lettuce, cucumbers, and tomatoes. Public health authorities in Hamburg first pointed the finger of blame at cucumbers from Spain, which they confidently claimed had tested positive for the responsible toxin. German federal authorities were more cautious, and their tests eventually concluded that those cucumbers did not contain the deadly bacterial strain, but by then Spanish vegetable growers had suffered severe damage to their reputation. Amid the confusion, Russia declared a monthlong moratorium in June on importing all fresh produce from the European Union.

About a month after the first German reports, a similar but much smaller outbreak occurred in a group of people who had eaten together at an event near the French city of Bordeaux. Detective work following that seemingly unrelated cluster of illnesses helped trace the cause back to sprouted fenugreek seeds from a particular organic farm in Bienenbüttel, a small town some fifty miles from Hamburg. The chain did not end there. Health authorities eventually concluded that fenugreek seeds imported from Egypt were the most likely culprits. The relevant bacterial strain was never actually isolated from the seed found at the Bienenbüttel farm. The European Food Safety Authority, asked to investigate the episode, sounded a note of equivocation:

The inability to demonstrate the presence of *E. coli* O104:H4 in the suspect seeds is not unexpected. It is possible that contaminated seeds were no longer in stock when sampling took place, or even if present were contaminated at a level which made isolation of the organism impossible. However, this does not mean that enterobacteriaceae would not have been present in seeds and sprouted seeds.[5]

In sum, even the combined expertise of Europe's foremost health research laboratories could not resolve all the uncertainties associated with a disease carried along on a supply chain that no single nation and no regulatory body could monitor in its entirety.

Together, the Rana Plaza collapse and the *E. coli* outbreak in Germany point to a recurrent problem of distributive justice in the global marketplace of risk. Those who bear the brunt of disasters are often those with least control over the circumstances that led to their injuries. These are people at the beginnings and ends of long chains of production: people working in dangerous factories, consumers of imported foodstuffs. They suffered or died because they had trusted intermediaries and higher-ups—employers, officials, experts, purveyors of goods—whose knowledge and judgment proved to be defective. Disasters, too, are fast-moving and they breed confusion. In unsettled and unsettling environments, blame can be wrongly assigned, causing yet more loss and grief. The cucumbers grown by Spanish vegetable farmers were eventually declared to be free of *E. coli*, but not soon enough to prevent a loss to Spain's economy of an estimated $8.5 million a day during the height of the crisis.[6] After disasters, help is often slow to arrive and inadequate when it comes, perpetuating a cycle of victimization. Mass poisoning episodes can induce chronic health effects that call for costly, long-term therapy. Legal procedures can tie up claims for years, partly because evidence collected during and after disasters is often unreliable and incomplete. Nothing, in any case, can adequately compensate elderly parents or small children for the loss of a child, a parent, or the family's breadwinner in chief.

An important step in the ethical governance of technology is to understand why these patterns of injustice persist around disasters and what could be done to alleviate their worst effects. For this purpose, three dimensions of disasters call for closer inspection: the limits of expert prediction; the constraints on compensation; and the sources of structural inequality in the management of technological systems. These three dimensions came sharply into focus in the world's worst industrial accident—the release of methyl isocyanate gas from a chemical plant owned by a subsidiary of Union Carbide Corporation in Bhopal, India, in 1984. Endlessly studied and endlessly debated, the Bhopal tragedy stubbornly resists any coherent narrative of causes and effects.[7] This makes the case particularly useful for probing the ethical anatomy of disasters. The narrative of Bhopal remains in this sense timeless. To be sure, it was something that happened a generation ago in a city few outsiders will ever visit, and yet in the conjunction of tragic failures that caused it, the Bhopal gas disaster became a parable for modern human overreaching and neglect.

DEATH AT MIDNIGHT

George Orwell turned the number 1984 into a metonym for dystopia in his account of a world in which an all-seeing authoritarian state crushes the spirit of individualism out of its citizens. The India of late 1984 was dystopic in quite another way, a rudderless state seemingly bereft of leadership and vision. Indira Gandhi, daughter of Jawaharlal Nehru, was four years into her third term as prime minister, a post from which she had been ignominiously ousted in the 1977 elections, a resounding referendum on her two years of authoritarian rule during the so-called

Emergency. Returned to power in 1980, Gandhi remained intolerant of dissent, yet unable to pull the country together. She quashed an incipient separatist movement by Sikhs of the northern state of Punjab with a bloody military attack on the Golden Temple, Sikhism's holiest shrine, in the city of Amritsar. Four months later, on October 31, 1984, Gandhi herself was assassinated by two of her Sikh bodyguards in revenge for the events in Amritsar. Violent anti-Sikh demonstrations in the following days, concentrated in the capital city of Delhi, led to thousands of deaths and the displacement of many more thousands.

The accident in Bhopal unfolded against that backdrop of bloodshed and political turmoil, but its roots were planted long before, in the history of India's struggle to achieve economic as well as political independence. Indira Gandhi's early years as prime minister coincided with the Green Revolution in India, a period when agricultural innovation saved the country from a humiliating dependence on foreign grain. Hailed as a scientific miracle, the seeds of the Green Revolution won the pioneering U.S. agronomist Norman Borlaug a well-deserved Nobel Prize. Seeds, however, cannot achieve their potential without intensive tending, and the short-stemmed, sturdy, high-yielding varieties that Borlaug bred were no exception. Mockingly referred to as "high-input varieties" by the Green Revolution's many critics, those lifesaving grains required enormous investments in water, fertilizers, pesticides, and electricity, whose combined costs and unequal distributive impacts are part of the history of this highly touted sociotechnical achievement. So is the disaster in Bhopal, which occurred at a plant tied to a system of agricultural production that increasingly depended on chemicals for its high yields.

In the late 1960s, Union Carbide Corporation (UCC), an American chemical company with its headquarters in Danbury, Connecticut, began operations in Bhopal, initially as a manufacturer of batteries. In 1974, UCC received permission to build a large new factory to produce one of its signature products, Sevin, the trade name for an insecticide containing the compound carbaryl. The parent company operated in India through a subsidiary, Union Carbide India, Limited (UCIL). Indian law at the time restricted foreign ownership of Indian concerns to below 50 percent, but because of its importance as a leading source of agrochemicals Union Carbide was exceptionally allowed to retain controlling ownership, with 50.9 percent of company stocks, while Indian financial institutions and private investors held the remaining 49.1 percent.

Carbaryl, unlike DDT, does not accumulate in animal tissues or the environment, and hence was considered an especially desirable pesticide for use on food crops. A lingering concern was that it does not differentiate well between harmful insects and beneficial ones, and the compound accordingly is not allowed for pesticide use in a number of countries. More important to the Bhopal story is that one of the ingredients used in manufacturing Sevin was a toxic intermediate called methyl isocyanate (MIC). The compound was known to be highly reactive with water, which converts liquid MIC into a gas. Even at extremely low doses, MIC causes intense irritation of eyes, nose, and throat, burns the skin, damages internal organs, and produces fluid buildup in the lungs, leading to coughing, suffocation, and, in many cases, death.

In the late evening hours of December 2, 1984, a large amount of water somehow penetrated into one of the underground storage tanks containing methyl isocyanate at the UCIL plant. Shortly after midnight, gaseous MIC began to leak into the air outside the factory. The scrubber system designed to neutralize the gas before it escaped was not working. The

alarm system that might have notified the community outside
the plant that an accident had occurred was not sounded till it
was too late. Within thirty minutes, a heavy fog of gas invaded
tens of thousands of homes in the densely settled, working-class
neighborhoods southeast of the plant. People sleeping in their
modest homes woke up choking, blinded, not knowing what
had engulfed them. Many died in their beds. Those who did
not struggled into the streets trying to run from the suffocating
fumes. A *New York Times* journalist reporting from Bhopal one
week later described the horrific scene:

> By the thousands, they stumbled into the streets, choking, vom-
> iting, sobbing burning tears, joining human stampedes fleeing
> the torment of mist that seemed to float everywhere. Some were
> run down by automobiles and trucks in the panic. Others fell,
> unable to go on, and died in the gutters along with water buffa-
> loes, dogs, goats and chickens.[8]

At least 2,000 people are thought to have died that night, and
another 1,500 soon after. Official estimates hover around 3,500
immediate deaths, though many believe the actual number
surpassed 8,000, citing the shrouds sold in the largely Muslim neighborhoods where the exposure occurred. As many
as 400,000 fled the city in panic, and estimates suggest that
150,000 to 200,000 suffered significant gas-related injuries. The
total number of the dead and injured remains contested to this
day, gradually rising as the long-term effects of MIC enveloped
more lives. What is painfully clear is that this was the largest
poison gas attack ever on a civilian population, and it occurred
outside the context of war.

Today, the site of the UCIL plant is deserted, enclosed by
wire fences and gates in an area where cows graze. UCIL no

longer exists as an entity; it was sold to McLeod Russell (India) Limited in 1994 and renamed Eveready Industries. Union Carbide itself is a skeleton of its former corporate self, a wholly owned subsidiary of Dow Chemical Company with some 2,400 employees, down to almost nothing from the 98,000 or so it employed at the time of the tragedy. Yet the names Bhopal and Carbide remain indissolubly linked in public memory, a memory so haunting that the now defunct company still takes pains on its public website to control and sanitize it.

ASYMMETRIES OF EXPERTISE

The UCC website provides a link to the company's history, a timeline of the rise and fall of a major multinational enterprise. The entry for 1984 reads as follows: "In December, a gas leak at a plant in Bhopal, India, caused by an act of sabotage, results in tragic loss of life."[9] This is the position Union Carbide has staunchly defended since shortly after the disaster, in legal proceedings, in public statements, and as part of its own self-understanding. According to the company, water could not have entered the plant accidentally, because the protective systems were working fine. It was the deliberate act of a "disgruntled employee," a person whose identity the company claimed to have ascertained as far back as 1986, and which the Indian government allegedly also knew but would not disclose, because that would have absolved UCC of legal liability. No actor in the Bhopal tragedy other than Union Carbide ever went on record to support the disgruntled employee theory.

Attributing the gas leak to sabotage neatly circumvented key questions about the knowledge gaps and power asymmetries that became visible in the disaster's wake. How much was known in advance, or not known, about MIC's long- and short-term effects on human health? Why was a substance of such toxicity stored onsite in a dense urban setting? What emergency measures were in place, and why were they so inadequate? Whose responsibility was it to care for the victims, not just in the accident's immediate aftermath but through long-term medical and social monitoring of chronic illnesses? UCC's threshold assertion that malign intent, not accident, had caused the disaster in effect absolved the company of any duty to address those questions in a public forum, although some answers can be gleaned from decades of investigation and legal activity that dogged the company in spite of its efforts to avoid culpability.

MIC was hard to study with standard techniques in the field of toxicology. Using animals, usually rats or mice, as surrogates for humans, toxicologists normally observe the effects of exposure to a target chemical for periods of up to two years, during which adverse health effects are supposed to appear. In 1984, very little was known about MIC, in part because toxicologists found the acrid compound so unpleasant to work with that they had not conducted studies to test it thoroughly. Most of what is now known about the chronic, or longer-term, effects of MIC (e.g., on lungs, reproductive systems, and mental health) thus derives from the worst possible kind of natural experiment: an unplanned exposure of unconsenting human subjects to a substance whose properties were not understood well enough in advance to allow for emergency preparedness, let alone long-term therapy.

A fierce medical controversy broke out in the days after the accident, convincing many Bhopal victims and their helpers that the company's and the state authorities' interest in evading responsibility had robbed gas-exposed people of a badly needed remedy. Disagreement centered on the distribution and use of sodium thiosulfate, a known antidote to cyanide poisoning. The authorities apparently feared a severe public relations backlash if the escaping gases were found to include hydrogen cyanide. Although victims said they were benefiting from thiosulfate injections, and some autopsy results seemed consistent with cyanide poisoning, state officials, backed by Union Carbide executives, denied those claims, insisting that such personal reports were untrustworthy and that, in any case, cyanide exposure had not occurred. The state of Madhya Pradesh summoned a police presence to shut down the dispensation of sodium thiosulfate in the people's clinics that had sprung up in Bhopal and even temporarily jailed some volunteer physicians working with the victims.

Mira Sadgopal, a physician at one such clinic, described the gulf between the victims' reported experiences and the position taken by the medical establishment:

> But people are talking abut distinct symptomatic relief for certain symptoms like headache, muscular pain, sleeplessness, chest pain, breathlessness, palpitations, blurring of vision—the responses are not uniform but certainly they are very distinct evidence of relief of the symptoms, but these results have not been accepted so far by almost all of the doctors in Bhopal, because they say there is no improvement in signs.[10]

On one side were the subjective feelings of poor and scientifically uninformed people; on the other, the expertise of orthodox medical science, including the respected Indian Council of Medical Research. There was little question in official minds as to whose views should prevail. Yet documents collected long after the fact, in legal discovery proceedings against UCC, indicate that reliable knowledge was in short supply.

UCC medical personnel who arrived on the scene shortly after December 3 were operating from a baseline of toxicological ignorance, not knowing in detail how MIC actually worked in human bodies or what might be the long-term, population-wide consequences of mass exposure. Minutes of a Chemical Manufacturers' Association (later American Chemistry Council) meeting on January 3, 1985, convened to discuss the events in Bhopal, offer disturbing confirmation: "the physicians sent to Bhopal after the accident are saying that the immediate physical problems for most of the survivors will probably disappear."[11] This optimistic assessment proved to be wholly unjustified, as diseases of the eyes, lungs, reproductive systems, and other organs, as well as psychological impacts, continued to accrue years after the disaster. In a battle between the embodied, experiential knowledge of victims and the speculative, unsupported claims of physicians, it is reasonable to think that the former should have received more credence. In practice, as the shutting down of the clinics dispensing thiosulfate showed, establishment medicine acting in the name of objective science displayed a callous disregard for victims' testimony, though there was little firm evidence to back up the official stance.

THE INJUSTICE OF LAW

If medical expertise did not serve the Bhopal victims well in their days of agony, then neither did the institutions of law, in India or the United States. A case such as this of alleged negligence leading to harm falls under the heading of torts, the branch of law that deals with injuries caused by private parties. Two serious practical problems immediately arose in the effort to apply tort law principles to the events in Bhopal. First, India at the time was a newly industrializing state and Indian courts were not prepared to deal with liability claims on the scale of what had happened in the wounded city. Second, foreshadowing by decades the events at the Rana Plaza or in Bienenbüttel, the causes and consequences of the Bhopal disaster crossed national boundaries. An American company's business, based on manufacturing expertise imported from the United States, had harmed untold thousands in India. Who could most effectively represent the uncounted mass of victims, whose claims ranged from urgent to questionable to possibly opportunistic, with the specter of many new claims yet to come as the disaster's full impacts unfolded? Which country should have jurisdiction over a continent-spanning case? And under what law should the plaintiffs' claims be heard and decided? Everything about the Bhopal litigation was unprecedented, and the twists and turns the case took over twenty years in addressing these questions deserve a monumental history of their own. Some aspects, however, stand out as especially salient in this examination of the disaster's ethical implications, aspects that transcend the particularities of Bhopal.

Almost as soon as news of the accident broke, a cluster of high-profile American trial lawyers with experience in mass tort litigation descended on the city in an effort to line up promising clients. American litigators were used to working for substantial contingency fees, that is, for minimal money up front but ultimately a quarter to a third of the damages awarded to plaintiffs at the end. In a case involving such egregious harm, it was tempting for U.S. tort lawyers to imagine that an award might set new monetary records. If the case were tried in the United States, Carbide's home jurisdiction, the victims could unite together in a single class action that would put considerable pressure on the company. There was every reason as well to hope for high punitive damages, and correspondingly high net gains for the plaintiffs' legal team. No wonder, then, that in those heady early days, some of the best-known names in the American litigation business buzzed around Bhopal like bees at a giant honeypot.

The Indian government stepped in to end what many saw as an unseemly example of transnational ambulance chasing. On March 29, four months after the accident, Parliament enacted the Bhopal Gas Leak Disaster (Processing of Claims) Act, 1985, known in short as the Bhopal Act. Building on the ancient common-law doctrine of *parens patriae* (parent of the country), which grants the state authority to protect subjects who cannot protect themselves, this law conferred on the Indian government the exclusive right to represent anyone who had made, or might in the future make, claims arising out of the events of that fatal December night. This right included the power to initiate and pursue lawsuits on behalf of the Bhopal claimants and also to reach a settlement of their claims. The law dashed the hopes of private lawyers hoping to reap big profits from the mass suffering in Bhopal, but it also pinned all of the victims' prospects on the Indian government's ability to represent their claims effectively and to secure a just and equitable legal outcome.

The Bhopal Act settled who was to represent the victims but not where. That issue had to be resolved on the basis of the *forum non conveniens* (inconvenient forum) doctrine, which holds that— in cases involving alternative jurisdictions—the suit should be tried in a place that does not pose undue hardships for the defendant or for witnesses. Ordinarily, in a tort action, this would be the jurisdiction where the injury occurred, unless that forum is shown to be inadequate. A strange pas de deux followed, in which the Government of India was forced to argue that the case should be tried in the United States because its own judicial system was incapable of dealing with a controversy of this magnitude, while Union Carbide counterargued that Indian courts were fully up to the task. Each side hired experts from the other country to help make its case on empirical grounds. India retained as its lead expert Marc Galanter, a law professor at the University of Wisconsin and the best-known American authority on the Indian legal system. UCC retained N. A. Palkhivala and J. B. Dadachandji, both seasoned barristers and senior advocates before the Indian Supreme Court. Palkhivala had served as Indian ambassador to the United States from 1977 to 1979.

It fell to Judge John F. Keenan of the U.S. District Court for the Southern District of New York to make the momentous ruling, and he ruled in Carbide's favor.[12] Important for Keenan was UCC's protestation that most of the documents and all of the contested behaviors and decisions were in India, along with the victims themselves, and it therefore made little sense to move the litigation to New York. This, the judge concluded, would place an unjustified burden on American judicial resources. Keenan also saw the case as entailing a politically sensitive verdict on India's sovereignty. Accordingly, on all points concerning the Indian legal system's ability to innovate, adapt, and create special procedures, Keenan rejected Galanter's pessimistic historical and sociological appraisals and accepted Carbide's assurances, backed by its distinguished Indian advocates, that India in a moment of crisis would rise to the challenge.

Toward the conclusion of his sixty-three-page opinion, Keenan offered a ringing paean to India's standing among nations:

> The Union of India is a world power in 1986, and its courts have the proven capacity to mete out fair and equal justice. To

deprive the Indian judiciary of this opportunity to stand tall
before the world and to pass judgment on behalf of its own people would be to revive a history of subservience and subjugation
from which India has emerged.

Ironically, by denying Bhopal plaintiffs access to the U.S. courts,
Keenan also denied them the benefits of innovations in the law,
such as class actions and contingency fees, that had considerably eased the path to redress for American mass tort victims.
On only two points did he make a concession to differences
between the two legal systems that might unduly prejudice the
plaintiffs' cause. He sent the consolidated cases back to India on
condition that Union Carbide subject itself to U.S. federal pre-
trial discovery rules, which allow plaintiffs considerably more
liberal access to records relevant to litigation, and that the company abide by any judgment rendered by an Indian court.

One significant attempt at innovation did occur within
the Indian legal system, and its fate in effect undercut Judge
Keenan's optimism. This was the doctrine of "multinational
enterprise liability," which holds that the parent company cannot delegate its duty to ensure that its activities do not cause
harm. Had it been accepted, this rule would have prevented
Carbide from claiming that fault, if any, for the Bhopal tragedy
lay entirely with UCIL, an autonomous corporate subsidiary,
and not at the doors of UCC's headquarters in Danbury. Indian
lawyers unofficially conceded that there was no precedent for
presuming enterprise liability in existing law, but took the position that the circumstances of this extraordinary case demanded
a new rule. Otherwise multinationals would always be able to
shield themselves against liability simply by creating a firewall
of subsidiaries between the home company and any hazardous
activities conducted in developing countries. Carbide's rejoinder

flatly rejected the proposed rule, citing the lack of precedents: "The defendant submits that there is no concept known to law as 'multinational corporation' or 'monolithic multinational.'"[13]

In the end, the validity of the multinational enterprise liability rule was never litigated. The chances always were that Carbide would settle the lawsuit instead of risking the expense and embarrassment of a protracted legal proceeding, not to mention the possibility of a crippling judgment. The questions were merely when, with respect to which claims, and, above all, for how much money the adversaries would strike a closing deal. The Indian government's asking price was above $3 billion, but in February 1989 the parties agreed to settle for $470 million—a paltry 15 percent of the initial demand that struck the 150,000 to 200,000 victims as absurdly inadequate compensation for all they had suffered and were still suffering. Nevertheless, the Indian Supreme Court upheld the legality of the settlement in December under the provisions of the Bhopal Act, thereby bringing to an end the most visible and debilitating of the controversies swirling around Bhopal.

Yet the exchange over the concept of enterprise liability demonstrated an almost unbridgeable gap between Judge Keenan's lofty hopes for Indian legal innovation and the real-world constraints of corporate and tort law that had evolved over centuries in developed countries to keep corporations safe from unexpected liability. The common law inherited from Britain by both India and the United States is not neutral vis-à-vis economic and political power. It is designed to deter revolution. Common law tends to conserve the status quo in its very forms of reasoning and in the ways it conceives of the role of the courts. Analogizing from one fact situation to another, and resisting speculation or arguments from pure principle, common-law adjudication is resolutely empirical, and as devoted to incrementalism as is British political culture, with its predilection for "muddling through." Common-law courts, moreover, are bound by precedent and rules of interpretation not to deviate too far from consensus positions about what the applicable law holds at any given moment. To stray beyond the limits of allowable interpretation, to be seen to be applying new rules, would draw courts into making policy, functions that common lawyers see as reserved for the legislative and executive branches, and hence as lying outside the prerogatives of judicial power.

India then was independent, but India was not free to choose its approach to litigation. Resistance to change was built into the foundations of the legal system within which Bhopal's citizens had to pursue their claims. Even the most basic rules of liability reflected conservative biases that favor the interests of accumulated property and organized capital. To win, the plaintiff's claim must be supported by a "preponderance of the evidence," that is, it must be more likely than not that the plaintiff's assertions are true. In Bhopal, legal rules and styles of reasoning, combined with scientific uncertainty, made the victims' case difficult to win within the framework of a conventional trial.

Chemical exposure cases present especially grave problems of evidence and proof because effects such as cancer or chronic diseases often appear long after the exposure, muddying causal claims, and because the precise amount and duration of exposure is seldom known with any certainty.[14] People in transient or outdoor jobs, such as farmworkers or gas station attendants, often are not monitored for routine workplace exposures; community exposure to toxic substances is even harder to establish with any precision. In postdisaster situations, high-quality evidence becomes still more difficult to generate: lines of responsibility are fluid and blurred, criteria for how to gather reliable data are not available, and disagreements abound. In Bhopal, the sodium thiosulfate controversy provided a foretaste of future problems as medical experts from different perspectives fought over what they were seeing and what should be done about it. And the sheer scale was overwhelming. No medical institution anywhere in the world could have coped with the burden of examining the hundreds of thousands affected by the gas cloud, or with the complexities of determining the nature, extent, and severity of their ongoing health complaints.

Transnational legal claims thus reveal a kind of imbalance that cannot be calculated through conventional accounting methods. At stake in the encounter between rich defendants and poor plaintiffs in multijurisdictional conflicts is the question whose idea of justice should prevail. Defendants, usually enjoying the economic upper hand, insist on stability and the application of well-entrenched norms, while plaintiffs see a need to revise the very rule structures that gave rise to their problems. That is an uphill struggle. A year or so after 1984, I organized a seminar on Bhopal at Cornell University, where I was teaching at the time. One of the invited speakers was a prominent attorney working for a major New York law firm representing Union Carbide. I asked her about the likely success of the "multinational enterprise liability" doctrine. Her reply was witheringly dismissive: "That has no basis in law." To her it was pure fiction, unworthy to be granted standing in expert legal thinking.

Formally, she may well have been right, though the answer cannot be known since the issue never went to trial. The deeper point, though, is that the novel doctrine of enterprise liability reflected a sense of justice warranted by actual circumstances, in this case an unprecedented experience of horror and loss, rather than by the law's established modes of expert reasoning. It was meant as a revolutionary response to a situation demanding drastically new ways of thought, and therefore as a self-conscious departure from rules of corporate liability codified long before hazardous enterprises went global. But law itself, as a practice internalized by experienced, Western-trained corporate lawyers, was not on the side of conceptualizing a new legal order committed, first and foremost, to justice. Corporate law in particular has a built-in bias toward stability, predictability, and the status quo, all of which favor the interests of capital. The demand for innovation, which Judge Keenan believed India to be capable of, had to pass through filters of logic and precedent largely designed to impede radical novelty. In a sense, then, the history of Bhopal in the courts can be seen as a struggle between law and justice, and law on the whole claimed the upper hand.

To be sure, the 1989 settlement did not extinguish the slew of collateral claims that continued to rise like a ghostly exhalation from the dead and abandoned site of the former UCIL plant. Prevented by law from seeking any direct adjudication of their primary claims, the Bhopal victims ingeniously pursued a variety of collateral actions, such as claims of bodily and property damage stemming from environmental contamination caused over the years by discharges from the Carbide factory. Their strategies became more sophisticated as the victims and their local advocates learned how to study their own condition and how to plead their case for themselves. But time was not on their side. Cases kept getting dismissed either as precluded by the 1989 settlement or as too old, too hard to prove, or not represented by appropriate claimants. Claimants died before their claims could be heard. In the aggregate, all of the collateral claims proved as inconclusive as the original suits arising from the 1984 disaster, though they ping-ponged their way through the U.S. federal courts for some twenty years.[15]

Law, it could be argued, does not always deliver timely or efficient justice even in wealthy nations. Hamlet complained of the "law's delay," and Charles Dickens used the tribulations of a never-ending lawsuit as the backbone for his 1852 masterpiece, *Bleak House.* It would be simplistic to deny, however, that despite all of its cumbersome qualities the law is more responsive and produces quicker results in the United States than in India. When the North Carolina–based Duke Energy spilled a massive amount of coal ash into the Dan River, polluting a seventy-mile stretch, federal prosecutors took immediate action to bring the company to account. A year or so later, Duke Energy agreed to pay $102 million in fines to clean up the mess caused by the leak. Later that same year, the company agreed to pay an additional $5.4 million to settle fifteen years of pending lawsuits under the Clean Air Act.[16] There was every indication that similar prosecutorial zeal will follow the Volkswagen's astonishing disclosure in September 2015 that it had used deceptive software to systematically avoid compliance with U.S. emission standards in its diesel-powered vehicles.

THE IMBALANCE OF POWER

Judge Keenan's faith in India's capacity "to stand tall before the world" reflects a confidence in the power of agency that some would say is characteristic of American social thought. It is founded on the notion that an actor endowed with will and imagination, especially one with the heft of a large nation-state, can get the results it wants through available channels of legal and political redress. The judge specifically noted the Indian courts' "proven capacity to mete out fair and equal justice." But fairness was not the issue in Bhopal. The question was whether Indian courts possessed the tools or the authority to force Carbide to compensate Bhopal's victims on a scale commensurate with the losses caused by its operations. To on-the-ground observers in India, the answer seemed obvious, and negative. In Indian eyes, the parity Keenan implicitly created between the Indian and the American legal systems was dangerously naïve, ignoring the sedimented structures of power that block those low down in the global hierarchy from effectively seeking redress for wrongs inflicted from above.

Most salient for Bhopal's mass of victims were the inequalities of bargaining power that produced the preconditions for the disaster. Carbide's majority ownership of the Bhopal facility—a fact that UCC lawyers dismissed as irrelevant because operational control, they argued, resided entirely with UCIL in Bhopal—stood for deeper dependencies that in the 1970s still plagued decolonizing nations such as India. Technological modernization, ardently embraced by Jawaharlal Nehru, India's first prime minister, shifted the lines of dependency without eradicating them. When India authorized Carbide to produce Sevin in the mid-1970s, the need to solve the nation's food security problems was visible and urgent, but India bought in effect a dangerously closed technological black box from Union Carbide. Importing a factory and a manufacturing process was not the same as acquiring full knowledge of the system's hazards or expertise in the associated practices of safety and management. Bhopal's sister plant in Institute, West Virginia, for example, was operated by workers trained for generations in the chemical industry. Even then, investigations conducted after the disaster revealed that there had been warning episodes at the Institute plant that might have sounded timely alerts for UCC's Indian venture—if anyone had known to ask the right questions in advance.

Inequality also infected the victims' relationships with the governments that were supposed to care for and, eventually, represent them in legal proceedings. Neither the Government of India nor the state of Madhya Pradesh, where the accident occurred, came to the postdisaster scene with altogether clean hands. Union Carbide designed the process and the factory for producing Sevin; and with criminal nonchalance Union Carbide allowed a still poorly tested technology, employing one of the chemical industry's most lethal compounds, to be imported into one of the world's most densely populated regions. But there was blame enough to spare for local governmental authorities who failed to inspect the plant and to heed earlier warning signs. That the scrubbers did not work and that no sirens sounded on the night of the disaster were failures of design, management, and enforcement. Yet it was not until after a tragedy of devastating proportions struck them that the citizens of Bhopal became aware of the Russian roulette the powers above had been playing with their lives.

It is sadly well-known that the consequences of disasters fall disproportionately on the weak and defenseless. Extreme conditions exacerbate entrenched inequalities, even in richer countries. Put differently, structure matters. In New Orleans after Hurricane Katrina and in Banda Aceh after the great tsunami of 2004, women, children, the old, and the poor died or were dispossessed in greater numbers than the rich and the able.[17] Bhopal was no exception. Fortuitously, the wind that night drove the gas toward the slum district, which also happened to be the city's poorer Muslim quarter, and the relative lack of turbulence caused the gas to settle at low levels, where people living in one-story dwellings typical of shanty towns were most prone to breathe it. Leaders of Bhopal's ongoing campaigns for justice and restitution remain convinced that the wheels of the law would have ground faster and more effectively had the wind on December 2, 1984, blown in a direction other than southeast.

Given the ordering of power, it is hardly surprising that the victims' deep desire to pin individual responsibility on company officials remained largely unmet. Days after the disaster, Carbide's CEO, Warren Anderson, flew to Bhopal, where he was arrested and briefly detained but then released on bail. He immediately returned to America and for the rest of his life never went back to a country where thousands continued to clamor for his trial and imprisonment, and where the courts in 1992 declared him a fugitive from justice. In June 2010, the Indian courts finally convicted seven aging former UCIL officials of homicide through negligence, a charge reduced from the original one of culpable homicide. To say that this was too little too late does not begin to capture the contempt and frustration of Bhopal's legal activists. Many likened what had happened in Bhopal as a crime comparable in significance to the events of September 11, 2001, in New York and Washington. For them the wildly uneven consequences of the two tragedies underscored all over again a story of inequality that they were powerless to reverse. Bhopal protesters frequently carried placards depicting Warren Anderson under a "Wanted" sign. In 2004, placards that I saw proclaimed, "You want Osama, Give us Anderson." But that demand went unheeded, like so many other pleas for acknowledgment from Bhopal.

CONCLUSION

In the aftermath of the Rana Plaza disaster, a nearly 500-page report authored by Mainuddin Khandaker of the Bangladeshi Ministry of Home Affairs found blame enough to go around among all of the principal actors: Sohel Rana himself, the archetype of the rapacious landlord, the so-called experts, the factory owners, the governments involved. But to reporters from the

German news magazine *Der Spiegel*, Khandaker put the matter more starkly: "That day, that April 24, was the inevitable result of the global market." Structure, in other words, matters more than agency, even though certain kinds of evil agency may be empowered within structures of injustice.

Technological disasters of the kind described in this chapter cast an unflattering light on the ethical contradictions that underpin many of modernity's prized economic and social achievements. As the world has drawn closer together, with denser global flows of money and commodities, those contradictions are often tragically evident. The first is the immeasurable discrepancy in knowledge and power that often separates the producers and the consumers of technological goods and services in a global market. Those at the receiving end—the workers in the Rana Plaza, the purchasers of sprouts in Germany, and the Carbide plant's neighbors in Bhopal—were never in a position to know the risks that lay just beyond the margins of their everyday existence. When things went wrong, their damaged and deceased bodies bore innocent witness to others' greed and lack of care. Helpless victims of the world risk society that Ulrich Beck passionately decried, they had no say in its design or processes of management.

A second feature of the global market that disasters bring to light is the asymmetry of the rules and norms that make markets function. In order to attract foreign capital, recipients of that largesse are continually driven to making concessions. Carbide's unusual majority ownership of the Bhopal factory and the Rana Plaza manufacturers' eagerness to get their employees back on the job in the face of dire warnings are examples of how unequally economic power works. Yet when accidents happen the victims must seek redress from the very same authorities that made the questionable regulatory calls and within the same structures of law that protect capital against demands for redistribution. A cry-ing sense of injustice, as the Bhopal litigation demonstrated, is not by itself enough to overcome the rigidities of the law.

Both law and the market imagine that social interactions can be played out on level playing fields, rendered equal for all with open information and fair rules of the game. This is a convenient fiction. There is in practice no level playing field on which to sort out claims whose causes originate in the built-in slopes and fissures of those fields themselves. What, then, can be learned from disasters? A renewed attentiveness to what I have called the "technologies of humility" is part of the answer.[18] These humble instruments include a heightened concern for the way technological risks are framed and measured, a focus on the needs of the most vulnerable, an assessment of technology's unequal distributive impacts, and a conscious effort to remember the mistakes of the past. The appropriate time for the application of these modest rules is before, not after, risky projects are undertaken. We will come back to these points at greater length in the concluding chapter.

Chapter 4

REMAKING NATURE

NEW PLANTS FOR OLD

In the long history of innovation, nature has served human beings in innumerable ways, as collaborator, resource, and sometimes reluctant sparring partner, to the point where scientists now question whether a nature independent of human influence even exists on Earth.[1] Agriculture emerged as one of the earliest instruments of large-scale human intervention into planetary dynamics. Wherever humans went, they foraged for food, eventually turning the full force of their experiential and theoretical knowledge to make the earth yield more of the commodities that serve their appetites and sustain their forms of social togetherness. The American historian William Cronon documented in his pathbreaking book *Nature's Metropolis* how the hinterlands serving Chicago changed beyond recognition as the city grew and had to be fed, turning prairie grass into cornfields, decimating herds of wild buffalo, and felling forests. Chicago, in its capacity to reconfigure nature, stands in for humanity's impact on the planet as whole. It is in this respect a microcosm, a miniature world.

From the time our hunter-gatherer ancestors settled down to

practice sedentary agriculture, they began selecting and breeding new varieties of plants and animals. Their impulse was utilitarian, how to get the best use from whatever was available. Farmers over the ages sought to produce higher yielding, better tasting, more attractive, and eventually less perishable varieties for increasingly far-flung markets. Plant breeding had its lighter side too, driven by aesthetics and the desire to produce new scents and colors to please the senses. Dutch tulip growers tried for centuries to breed the elusive perfectly black tulip. More recently, Japan's Suntory brewing company funded a competition to create a blue rose through genetic engineering; years of effort resulted in a flower that strikes dispassionate observers as more lavender than true blue. Fruit and vegetable growers continue to turn out untold hybrid varieties, from seedless watermelons to exotic creations such as the pluot, a cross between a plum and an apricot, and the tangelo, hybridizing the orange with the pomelo.

Given this long history of manipulating nature, the agriculture industry of the late twentieth century saw nothing wrong with modernizing plant breeding and animal husbandry with the new techniques of genetic engineering. In the United States, farmers and growers had happily collaborated with university-trained agricultural scientists, marrying research and application, since the middle of the nineteenth century.[2] Now the sophisticated forms of genetic modification developed in academic research labs promised a cornucopia of new breeds of genetically modified organisms (GMOs) with commercially valuable characteristics, such as improved yields, novel therapeutic and nutritional properties, and resistance to extreme environmental stresses. By moving genes between species, scientists could create transgenic strains that would not have arisen naturally, through normal evolutionary processes. These properties could help plants tackle some of nature's most intransigent threats, from pests to drought and other consequences of climate change. In spite of its grand promises, however, agricultural biotechnology turned out to be one of the most controversial technological applications of recent times, a paradigm case for what to avoid in introducing wide-ranging transformations into modern industrial production.

Why did a technique that seemed so appealing—ingenious, feasible, broadly applicable, of huge potential benefit to the poor and hungry, and commercially profitable—kindle an ethical and political firestorm that refuses to die down? To make sense of this puzzle, "green biotechnology" has to be understood in the context of the political economy, indeed the multiple political economies, of global agricultural production at the millennium. The debates around this technological development evolved, we will see, out of the same overlapping histories of invention, market expansion, and global circulation that led to the Bhopal disaster described in the preceding chapter.

Discoveries originating in the lab may appear value-neutral to bench scientists, but they raise thorny issues of governance when they make their way into the field and into commerce. Unexpected conflicts may arise as the scientific and technological achievements of advanced industrial nations are disseminated to other regions of the world, where they bump up against radically different farming practices and consumer preferences concerning risk, safety, and the value of nature. Such considerations were not at the forefront when U.S. scientists first worked out the techniques of genetic modification. University scientists, industry, and policy institutions all assumed that GMOs would be accepted as an unquestioned benefit so long as they posed no appreciable risks to health and safety. Risk assessment, moreover, was a job for experts who understood biological processes. Consumers on the whole were not consulted about their preferences in the early stages of product development, though they proved after the fact to have strong and justifiable opinions about what they ate and how it was grown. No amount of policy catch-up has compensated for those early shortcomings of public deliberation.

The story of agricultural biotechnology at the turn of the twenty-first century concerns in large part the inability of existing institutions to mediate between the hopes of ambitious technology promoters and the doubts and fears of people who see biotechnology as an ungovernable force and a bearer of disruptive cultural values. With regard to benefits, many have questioned whether the product and process innovations pursued by a fledgling industry were best suited to serve the world's neediest. Others have asked whether risks were fully and dispassionately assessed by the profit-hungry corporations driving technology development. Can self-interested manufacturers responsibly calculate the risks of creating unruly organisms that may escape and overwhelm natural biodiversity or turn poisonous to humans and their environments? Is the talk of risks and benefits itself a thin veneer for a sociotechnical transformation that threatens traditional lifestyles and global food security? These questions have been on the table for decades, but the answers remain as far apart as the underlying political positions.[3]

STRATEGIC ERRORS

Exaggerated claims and strategic missteps dogged agricultural biotechnology from its very beginnings. Innovation was guided more by what industry could imagine and implement than by growers' needs, let alone by what customers wanted to buy or eat.

Several early ideas proved politically disruptive and commercially unviable. The first U.S. initiative was not even a plant; it was the so-called Ice Minus, a variant of the common bacterium *Pseudomonas syringae* (*P. syringae*) with one gene deleted so as to reduce its ability to form ice crystals on its surface. Proponents hoped that spraying the gene-deleted bacterium on vulnerable plants, such as strawberries, would protect them against sudden frost and lowered yield. They noted that the Ice Minus strain of *P. syringae* already occurs as a mutant in nature and hence is not a new organism, but that argument failed to satisfy opponents. Environmentalists worried that large-scale field applications of Ice Minus would upset ecosystem balance and might have deleterious consequences that no one had carefully studied. A consensus eventually formed that it was premature to release genetically engineered microorganisms into the environment. Industry attention shifted instead to modifying crop plants and farm animals so as to improve their commercial viability.

Those efforts, too, ran into choppy waters. The unhappy fate of the Flavr-Savr tomato offers one example. In 1980, agricultural scientists from the University of California at Davis formed a company, Calgene, to carry some of their ideas from the lab to the field. Their plan was to engineer a tomato that would stay longer on the vine, develop fuller flavor, and still be transportable to distant markets. Trade-named Flavr-Savr, the tomatoes were genetically modified to produce less of an enzyme that softens cell walls as the fruit ripens. Calgene scientists succeeded in their primary aim, but the Flavr-Savr proved to be a commercial disappointment. Slower ripening did not enhance flavor as the scientists had hoped. Nor did the product travel well. Flavr-Savr tomatoes tasted too bland to satisfy consumers, and high production and distribution costs meant they could not outcompete conventional tomatoes. For some years,

Flavr-Savr tomatoes from California were processed into tomato paste and distributed by large British supermarket chains, but with rising public resistance toward all forms of GM crops in Britain, that market, too, dried up by the end of the 1990s. Calgene followed the life cycle of many successful biotech startups. It was acquired by Monsanto in 1996.

Early industry efforts to use GM techniques in animal husbandry proved no more satisfactory than the Ice Minus and the Flavr-Savr. Once again, the lure of the technically possible led to false starts and dismaying turns in relations between the biotech industry and the wider public. A U.S. government–sponsored project for inserting the gene for human growth hormone into pigs produced animals suffering from arthritis and other diseases; notoriously, seventeen of the nineteen "Beltsville pigs" died within a year. Similarly, the use of genetically engineered bovine growth hormone (rBGH) to increase milk yield in dairy cows caused painful inflammations of the udder and other forms of physical distress in treated animals. In turn, these infections led to increased antibiotic use and, many feared, the threat of antibiotic resistant bacteria down the line. Economic factors, too, came into play, since rBGH use favored larger farmers who could benefit most from yield increases. Regulatory agencies in the European Union (EU) and other countries took these consequences seriously enough to ban rBGH in their dairy industries. The U.S. Food and Drug Administration (FDA), however, construed its responsibilities more narrowly. As a body principally responsible for human health, FDA approved rBGH use on the ground that milk from treated animals poses no risk to humans. The government's failure to regulate outraged small farmers and animal welfare groups and propelled a demand for organic dairy products free from traces of rBGH. Many American consumers still refuse to buy milk or milk products derived from cows treated with rBGH and regard the hormone's spread as a prime example of industry's heedless use of biotechnology. Yet, under existing law and policy, no U.S. federal agency has unambiguous authority to consider these ethical and economic concerns or even the full range of ecological uncertainties associated with the global spread of GM products.[4]

Despite some early setbacks, U.S. agricultural biotechnology made huge strides in the 1990s through an alliance between manufacturers and commodity crop growers wishing to combat specific plant pests, such as the bollworm or the European corn borer, that are responsible for major losses in production. Growers especially welcomed a form of genetic engineering that inserts a gene from a bacterium known as *Bacillus thuringiensis* (Bt) into crop plants. The Bt gene produces a protein that kills the larvae of harmful insects and yet is considered harmless to human beings and beneficial insects such as bees and butterflies. A second commercial success was the production of herbicide-tolerant crops through genetic alterations that make these plants resistant to a common form of weed killer. This invention turned the U.S. chemical giant Monsanto into the global face of agricultural biotechnology. Monsanto manufactures a widely used herbicide named Roundup, based on the chemical compound glyphosate. GM technology allowed Monsanto to produce a range of "Roundup-Ready" plants that could be marketed as a package with its already popular herbicide. Farmers could safely spray their fields with Roundup during the growing season, knowing that the herbicide would kill only the undesirable weeds and not the cash crops. Roundup remains extremely popular with growers, although some concerns have been raised, on the basis of inconclusive studies, about glyphosate's possible carcinogenicity to humans.[5]

In less than twenty years from their first introduction around

1996, the percentage of GM soybean, corn, and cotton soared in the United States. Based on its surveys, the U.S. Department of Agriculture estimated that herbicide-tolerant soybeans accounted for 94 percent of all soybeans grown in 2015, while the corresponding figures for herbicide-tolerant cotton and corn were 89 percent that year for both. The figures for insect-resistant Bt cotton and corn were 84 percent and 81 percent, respectively.[6] These numbers far exceeded global average uses of GM crops, attesting to a particularly comfortable alliance between the U.S. biotech industry and American commodity crop producers.

NOT EVERYTHING IN THE GARDEN IS ROSY

Buoyed by successes at home, Monsanto and other American companies engaged in GM crop production turned their attention to the global market in the 1990s. Here, however, they encountered a fierce onslaught of negative publicity and public rejection that caught them wholly off guard. Remarkably, trouble began in Europe, though most European governments and EU regulators in Brussels were very favorably disposed toward a technology that they saw as beneficial and market expanding. Publics, however, thought otherwise. European environmentalists and small farmers led the charge against a technological invasion from America that many viewed as untested, unnecessary, and unsustainable. Debate polarized across the Atlantic. American producers and their political allies bemoaned Europeans' seemingly uninformed and unscientific judgments, while Europeans contested U.S. claims of safety, arguing that these were often based on ignorance and guesswork rather than on reliable science. Monsanto became the lightning rod for much that seemed wrong with American attitudes toward the rest of the world at the millennium—arrogant, culturally insensitive, and unwilling to concede the limits of its own scientific knowledge and technological capability.

The great GMO debates of the turn of the century revealed many gaps in science's understanding of the way information encoded in the genes functions within complex organisms, let alone in larger ecological environments. Importantly for our purposes, they also highlighted the weaknesses of institutions responsible for governing emerging, globally diffused technologies. When facts are unknown or unclear, who is ultimately responsible: for generating missing knowledge; for striking a balance between precaution and risk taking; for mediating between conflicting ideas of nature and sustainability; and for picking up the costs when accidents happen? In each case, the answers remain elusive or deeply problematic from the standpoint of good governance, on national as well as global scales.

"Unknown Unknowns"

Oddly enough, a controversial U.S. political figure supplied language that was widely picked up in framing the policy issues around GMOs. Donald Rumsfeld, twenty-first U.S. secretary of defense, under President George W. Bush, left two legacies when he stepped down from his last public office in 2006: first, the corrosive economic and political consequences of the "war against terror" in Iraq and Afghanistan; second, an instantly popular phrase that sums up the contradictions of all purposeful action in modernity. That phrase was "unknown unknowns." In a statement of haiku-like simplicity, Rumsfeld said at a 2002 press briefing:

There are known knowns.

There are things we know we know.

We also know there are known unknowns.

That is to say, we know there are some things we do not know.

But there are also unknown unknowns—

The ones we don't know we don't know.

This "brilliantly pithy piece of popular epistemology,"[7] as one observer called it, caught on to such an extent that Rumsfeld adapted it for the title of his 2011 memoir, *Known and Unknown*. The famed documentary filmmaker Erroll Morris, a lover of mysteries, went one step further, naming his 2014 documentary on Rumsfeld after the only combination the secretary left unlabeled, *The Unknown Known*. Clearly, the concept of unknown unknowns struck deep chords of anxiety in a society that depends on knowledge as the foundation for any kind of decisive action.

Risk assessment, as described in chapter 2, deals largely with known unknowns. It is an effort to bound and quantify the probabilities of outcomes that are already within the human imagination, and thus in the realm of the knowable. In the earliest years of genetic engineering, that framework was seen as adequate for responsible regulation. Leading molecular biologists met in 1975 at California's famed Asilomar Conference Center to discuss how GMOs, including dangerous pathogens never before seen in nature, might be kept from accidentally escaping from the laboratory and harming human health or the environment. The conferees knew from their own lab work that accidental releases happen, and they tried through policy to contain the predictable risks of a known hazard. The result was a system of physical and biological controls on lab research to ensure that the terrifying possibilities imagined at Asilomar would never come to fruition.

At that time, however, recombinant DNA technology was still like a toy in the hands of eager children. Few of the molecular biologists assembled at Asilomar anticipated the rise of whole new industries built on intentional releases of GMOs into the environment. They did not imagine that genetically modified varieties of staple crops such as corn or cotton would one day almost entirely supplant naturally occurring varieties. Quite simply, that distinguished body of scientists focused on the world they knew best—research labs, particularly ones oriented to biomedical research; commercialization was not yet their center of attention. Dangers arising from "deliberate release," that is, intentional, large-scale introductions of herbicide or insecticide-resistant crops, were, in Secretary Rumsfeld's terms, "unknown unknowns." Yet, within just a few years, that prospect was very much a reality, championed by industry leaders such as Monsanto and Syngenta that were already heavily involved in agricultural technologies before GM appeared on the stage of possibility.

Knowledge about the risks of deliberate release accumulated slowly and patchily, more through post hoc experience than by prior assessment. A series of accidents in working with GMOs sheds light on the gaps and holes in the dominant U.S. imagination of governable risk. In 2000, a strain of Bt corn known as StarLink, approved for use in animal feed but not for human consumption, turned up in corn products made by Kraft Foods and sold by the fast-food chain Taco Bell. The GM variety contained a protein called Cry9C, which the U.S. Environmental Protection Agency (EPA) had classified as a potential human allergen. Genetic ID, a lab working with a consortium

of several anti-GMO environmental and consumer organizations, conducted tests on Taco Bell products and identified a
contamination that should never have happened. Subsequent
buybacks, product recalls, and loss of exports ran into hundreds of millions of dollars, and StarLink itself was taken off
the market.[8] One study estimated that the episode depressed
corn prices by close to 7 percent for at least one year.[9] The case
was nonetheless widely dismissed as an avoidable accident—an
"unintended consequence" of fundamentally well-intentioned
actions. The EPA, according to one argument, should not have
approved StarLink only for animal feed because the agency
should have known that it is difficult to segregate approved
corn varieties from nonapproved ones in a complex manufacturing environment.[10] Alternatively, if grain handlers had only
been more careful and followed the rules, the mix-up would
not have occurred. Besides, some said, the scare was out of all
proportion to the very few documented cases of allergic reactions actually reported to the Centers for Disease Control. In
short, for biotech advocates, StarLink revealed nothing intrinsically wrong with the process of plant genetic modification.
The case represented only an unfortunate deviation from good
regulatory and manufacturing practices, compounded by public overreaction.

Yet the StarLink episode also underscores that, when it comes
to new technologies, imagined futures do not correspond well to
the institutional realities of knowledge flow and responsibility.
We saw in chapter 2 that nonknowledge, in particular, is not an
absolute but a relative category: what one knows depends crucially on one's standpoint, especially one's place within an organization, a process, or a hierarchy. What a geneticist or medical
scientist knows about human allergies is vastly different from
what the operator of a grain elevator knows about conditions

of seed storage and shipment. These multiple forms of knowledge are typically segregated from one another in the course of industrial processing. Tellingly, abstract scientific knowledge tends to be privileged by risk assessors. Pragmatic knowledge of the kind processed by silo operators rarely finds its way into peer-reviewed scientific articles or the rarefied forums of risk assessment and policymaking that rely on published science. The resulting official picture of risk or safety across a complex technological system may therefore be misleadingly partial and incomplete.

"Unknown to whom?" becomes, then, an important threshold question, and ill-judged answers can be ethically troubling as well as harmful for society. Scientific and technological unknowns may seem unknown only because the most authoritative knowers lack perspectives that might have been available from less elevated points of view. Thus, the molecular biologists who dominated the early regulatory discourse on GMOs were most sensitive to risk from accidental releases of the kind they could imagine. They were not as knowledgeable, or indeed as concerned, as ecologists later came to be about the harmful impacts of GMOs on nontarget species, the appearance of resistant traits, or the risks of gene transfer between modified and nonmodified organisms. Failures of monitoring and control in messy production systems were even more foreign to the high-minded Asilomar scientists. Those concerns emerged only later, in conflicts among experts from different fields and between the U.S. biotech industry and reluctant farmers or consumers in other countries. Cross-national differences in expert assessments and control philosophies loomed into view only after U.S. regulatory strategies based on relatively narrow understandings of risk were already in place.[11] U.S. manufacturers, however, accepted their own domestic experience of regulatory risk assessment as the gold standard for sound science and dismissed criticism from other countries as not scientific.

The regulatory history of GMOs illustrates how those in positions of power, whether scientific or political, lack incentives to seek out viewpoints contrary to their own when assessing and managing risks. Claims arriving from outside the dominant frameworks of thought frequently are rejected as ignorant, unfounded, or lacking scientific validity. The phenomenon of "unknown unknowns" can be seen in this light as more a statement about the unequal distribution of power—the power to determine whose knowledge counts and for what purpose—than about an absolute vacuum of relevant knowledge.

Risk or Precaution?

Since knowing is partly a matter of perspective, and complete knowledge is an unattainable ideal, the globalizing world is full of situations in which one group's claim to know things is denied or contradicted by another group claiming better knowledge or good reasons for doubt. International trade in GMOs illustrates one such protracted standoff, underlining again the inadequacy of existing global institutions to mediate between sharply divergent views on what needs to be known and how to cope with uncertainty and value conflicts.

By the early 2000s, a major rift had developed between nations and organizations committed to a "science-based" or "risk-based" approach to dealing with agricultural GMOs and those who favored "precaution."[12] Those terms have been written into law and policy in varied ways, and none has a single straightforward definition, but there is a consistent difference in what each connotes about the limits of scientific knowledge. Broadly speaking, science-based (or risk-based) approaches frame problems as known unknowns, and assume that reliable predictions about the future can be made on the basis of current knowledge or targeted further studies. By contrast, precautionary approaches take for granted that some important things lie beyond the limits of scientific inquiry, including the entire category of unknown unknowns. How can one assign probability estimates, as risk assessors seek to do, to a problem that one cannot even imagine? More science is not necessarily a good answer when a technology's future trajectory presents serious uncertainties. Indeed, by generating more knowledge about what is very poorly understood, science could even distract attention from better-known problems that happen to lie outside the perimeter of scientists' conventional wisdom.

The clash between risk and precaution erupted into a transatlantic trade war on GMOs. In 2003, the United States, Argentina, and Canada—all big grain exporters—began a case against the European Union in the World Trade Organization (WTO). The EU, the complainants charged, had violated international rules of free trade by maintaining an illegal moratorium against the importation of GM crops. Refusal to admit these crops was unlawful because risk assessment had shown them to be safe and because the EU had no valid scientific basis for keeping GMOs out. None of the risk assessments on GMOs offered any suggestion that they cause harm to human health or the environment. Accordingly, it was unlawful to erect barriers against these products without countervailing scientific evidence of risk. Such unsupported action flew in the face of the free-trade principles that all WTO member countries subscribe to.

Some two years later, the WTO dispute settlement process concluded in a 1,000-page opinion that the EU had indeed maintained an illegal moratorium against GMOs by causing "undue delay" in their approval. Delighted, the biotechnology industry touted the decision as vindicating its claims regarding crop safety, but reaction among European consumers and some EU member states was a lot less friendly. This was a case in which American risk assessments, based on what many saw as an inadequate examination of ecological and possibly even public health unknowns, seemed to be trumping other nations' legitimate doubts. WTO's reasoning acknowledged unresolved tensions between free trade and political judgments about how much and which kinds of uncertainty can be tolerated when nations buy each other's technological products and, implicitly, their safety standards. The case sharpened the conflict between the science-based and the precautionary worldviews. As a formal legal matter, science prevailed, but despite the power of legal affirmation, doubts about the universal claims of U.S. science simmered and refused to vanish.

From the WTO's point of view, the dispute centered on the interpretation of key treaty language, especially articles 5.1 and 5.2, which set forth the basis for any trade-restrictive measures that nations choose to adopt. In particular, article 5.1 explicitly endorses the science-based approach to regulating cross-border movements of technology:

> Members shall ensure that their sanitary or phytosanitary measures *are based on an assessment*, as appropriate to the circumstances, *of the risks to human, animal or plant life or health*, taking into account risk assessment techniques developed by the relevant international organizations. (emphasis added)

Article 5.2 further provides that risk assessments must be based on "available scientific evidence," while article 5.7 insists that, in cases where nations believe the science is insufficient, they have a duty "to obtain the additional information necessary for a more objective assessment of risk."[13] Together, these provisions strongly endorse the view that science, either extant or practically obtainable, can resolve the uncertainties germane to trade in GMOs. There is little room here for the possibility that some sorts of uncertainty cannot be eliminated through further research: because nature is too complex; because we cannot imagine all possible causal pathways; or because the studies that would be required are prohibitively time-consuming or too expensive to undertake.

The commitment to science as a supposedly neutral arbiter of trade disputes puts the WTO in an awkward position. Is it applying objective principles of science and law, or is it impermissibly intruding on national sovereignty? If risk assessment is itself a value-laden means of managing uncertainty, then sovereign states may choose to adopt different approaches if their publics demand greater precaution in the face of unknown unknowns.[14] In choosing between risk-based and precautionary stances, then, the WTO inserts itself into the space of national sovereignty by choosing between different "philosophies of regulation."[15] The WTO acquired that position of supersovereignty, however, almost by the way, as an unwritten corollary of its treaty-enforcing powers, without any semblance of explicit prior deliberation on whether signatory states were willing to abide by the WTO's epistemological preferences.

A Taste for Nature

The science-based approach to uncertainty written into the WTO treaties rules out of bounds the value conflicts that invariably accompany attempts to modify or commodify nature. Many people distrust genetic modification because they see it as moving too fast and too far beyond the bounds of traditional farming and breeding techniques. Skeptics include, among others, small farmers, locavores, growers of specialty or heritage crops, and ethically committed buyers of organic food. All these groups have a strong desire not to see industrialized agriculture, including GM technologies, displace their preferred ways of growing and consuming food. On both sides of the Atlantic, and globally, markets and supply chains have sprung up to cater to a growing consumer preference for organic agriculture, with scrupulous attempts to keep the organic apart from the nonorganic. A coffee grinder in a Whole Foods supermarket in Cambridge, Massachusetts, for example, bears the following sign: "This grinder is used for both organic and conventionally grown coffee beans. Those customers who are passionate about maintaining the organic integrity of their coffee may want to consider grinding their beans at home."

That sign, despite its somewhat tongue-in-cheek formulation, reflects the storewide policy of "full GM transparency" that Whole Foods Market adopted in March 2013. The company announced a five-year deadline within which it would label all products in its North American stores to show whether they contain GM ingredients. Private initiative in this case ran way ahead of governmental response. Whole Foods' gesture contrasts markedly with consumers' failure to get similar action from the federal government for food sales across the board. Indeed, U.S. policy actively discourages GM labeling. In one famous test case, the FDA and the U.S. Federal Trade Commission (FTC) warned specialty dairy product manufacturers, such as the popular ice cream maker Ben & Jerry's, that they would be found guilty of false and deceptive advertising if they certified their products as free of genetically engineered recombinant bovine growth hormone. Organic dairy product manufacturers were therefore obliged to follow a more contorted route, claiming only that they source their milk from farmers who guarantee that their herds are not treated with rBGH. Even then, the labels must also declare that the FDA has found no significant difference in milk from rBGH-treated and untreated cows.

Farmers who opt against genetic engineering are caught between increasing numbers of consumers who say they want organic products and a powerful industry lobby whose products drive down prices and who are willing to spend a great deal of money to keep the non-GM market niche from expanding. In the United States, state laws mandating GMO labeling have become a significant battleground, extending well beyond the relatively contained territory of rBGH use in the dairy industry. A failed 2012 statewide referendum in California, Proposition 37, that would have made GMO labeling a legal requirement aroused intense passions on both sides. A late but impressively funded attack on the referendum led to its narrow defeat, with about 51.4 percent voting against it. The biotech and food industries, spearheaded by Monsanto and DuPont, vastly outspent the measure's supporters, raising some $46 million against the proponents' $9 million. A similar effort lost by a somewhat wider margin in Washington State in 2013. The fight, however, continues with bills to require labeling pending in several state legislatures.

The desire to keep industrial and nonindustrial agricultures apart is not limited to consumers wanting a more natural diet for themselves or their families. Organic growers recognize that the market for their products depends on reassuring buyers that their crops are not contaminated with GM ingredients.

National governments, too, have an interest in ensuring that choice economic sectors, such as organic producers, do not incur insupportable losses through accidental contamination of their products. Strict rules of "coexistence," such as those adopted by the European Union, seek to ensure that GM products will be traceable—that is, identifiable through signature markers. Recognizing that total purity is unattainable, EU coexistence rules provide that a product may be designated GM-free if it contains less than 0.9 percent GM ingredients; Australia and New Zealand use a rounded-off 1 percent limit. These numbers are considered practically feasible, though even lower levels could be detected with existing technology. In the United States, the FDA continues to recommend against the GM-free label because, in the agency's opinion, that designation misleadingly suggests 100 percent purity.

Disorderly nature and neat bureaucratic logic, however, do not make good bedfellows. European regulators discovered this when they tried to do an end-run around the debate over GMOs with the principle of coexistence. Based on respect for free consumer choice, that principle signals a commitment that all forms of agriculture should be given a chance to thrive in Europe. To implement coexistence, EU authorities prescribed distances that must be maintained between areas planted with modified crops and unmodified ones. Those distances are normally adequate to prevent wind-borne pollen transfer,[16] but they turned out to be insufficient to keep bees from flying between GM and non-GM plots, foraging on pollen from both kinds of plants. The honey from those undisciplined bees may contain more than the 0.9 percent GM ceiling for the GM-free designation. Beekeepers selling such honey to determined consumers of natural products face a loss of income. Europe's small beekeepers, moreover, see coexistence as intrinsically harmful to a form of agricultural life that depends on itinerant farmers freely moving their colonies around the country in search of the best pollen sources. Constrained and threatened by what they saw as an overly permissive regulatory regime, German beekeepers mobilized in Munich 2008 to demand political asylum for bees that they claimed had been driven off their land.[17]

On a grander political scale, several European countries declared bans on some or all forms of GM cultivation. Targets include, most prominently, Monsanto's insect-resistant hybrid corn MON810, one of the few GM species approved for planting in Europe. The company claims that these bans are illegal, on the basis, for example, of a 2011 decision by the European Court of Justice declaring illegal a French ban on the cultivation of MON810.[18] Flouting court decisions, France subsequently renewed the ban, most recently ahead of the 2014 growing season. Monsanto also cites the failure of any health and safety agency, including the European Food Safety Authority, to find any risks of injury to humans, animals, or nontarget species.[19] These arguments, however, merely underscore the unbridgeable gulf in values between an industry that believes in risk-based approaches to regulation and a consumer and grower alliance that fears unknown unknowns and has opted for different, more locally grounded modes of agricultural production. A discussion framed solely in terms of physical and biological safety does little to address worries about a dangerous concentration of power over plant and seed varieties, and correspondingly reduced variety in the food supply, through GM manufacture controlled by a handful of large multinationals.

Who Pays?

When outcomes are uncertain, people may be reassured to know that someone will pick up the costs if the unthinkable happens. This is why we insure ourselves against fires, floods, car accidents, and other common hazards of modern life. When harm ensues from breakdowns in large technological systems, however, the rules for who should pay and for what kinds of damage, especially in cases of low-probability, high-consequence events, are not always clearly spelled out in advance. Ramifications are often unpredictable, as dramatically illustrated by the StarLink episode. The discovery of Cry9C in commercially made taco shells necessitated a costly recall that eventually reached hundreds of products, affected companies in the United States and Mexico, and threatened U.S. exports to Japan and South Korea. To protect domestic corn growers, the U.S. Department of Agriculture spent $15 to $20 million to buy back unused StarLink seed. Aventis, the Franco-German company whose subsidiary had marketed StarLink, agreed in 2003 to pay $110 million to farmers hurt by the drop in corn prices and more than $500 million in aggregate payouts to others affected by the recalls.[20] The total cost of cleaning up the corn supply after StarLink has been estimated to be as much as $1 billion. These costs diffused through society without any precise reckoning of how they were allocated between Aventis and wholly innocent public or private actors.

A different, more tragic story of payment and responsibility unfolded around the introduction of GM cotton to India in the early twenty-first century. Trade liberalization opened up India's markets to foreign investment in the 1990s. That policy change paved the way for multinational agribusiness companies to begin selling their seeds in a country with the world's highest concentration of small farmers, who had previously subsisted on traditional seeds, replanting their own saved seed from year to year. Monsanto's Bt cotton, trade named Bollgard, seemed a godsend for India's cotton farmers, plagued by bollworm infestations that devastated their crops. After three years of trials, the Indian government approved Bt cotton for commercial use in 2002, and within ten years it accounted for some 95 percent of India's total cotton crop. The conquest of the bollworm promised higher yields, with more stable profits for farmers, combined with lower use of chemical pesticides and less harm to biologically beneficial insects and plants. The global biotechnology industry, in short, promised a win-win scenario for Indian cotton manufacturers and users, with big payoffs for the Indian economy. Bt cotton's near-complete takeover from non-GM precursors in just one decade, paralleling developments in the United States, seemed to fulfill that promise.

Already by the late 1990s, however, stories began to circulate about disproportionately high suicide rates among Indian farmers. Driven to the wall by dropping prices and rising costs, uncounted numbers chose death by hanging or by swallowing poisonous chemical pesticides as the only way out. Death released the debtor but did not extinguish the debts, and distraught families were left to bear the burden.[21] The national government and some states with especially high rates of farmer suicide enacted debt relief laws to help distressed farmers and their families, but coverage was rarely adequate to prevent tragedy.

The coincidence of economic liberalization, the government's approval of GM cotton, and reported farmer deaths caused consternation among activists worried about foreign monopolies controlling Indian agriculture. One line of argument pointed to GM cotton as a specific cause of increased suicides. Advocates of this view noted that small farmers who had previously paid little or nothing for their seeds were now forced to buy expensive packages of seeds, pesticides, and fertilizers, cutting into their potential profit margins and leaving them vulnerable to bad harvests and growing debt. The biotechnology industry countered with a variety of arguments, from blaming the deaths on alcoholism or overspending on children's education to citing studies that showed no significant change in farmer suicide rates after the introduction of GM cotton.

Predictably perhaps, the debate turned into dueling statistics, with each side attempting to discredit the causal connections, or lack of connection, that seemed so plain to the other. This was not simply a matter of extremists toeing hard-line positions and selectively citing data to fit their cause. Rather, polarization even among mainstream researchers indicated that here was a matter falling in that gray zone of complexity, where facts and values are so tightly coupled that neutral determinations are essentially impossible to come by. In March 2014, the *Economist*, a voice of respectable probusiness opinion, approvingly cited on its *Feast and Famine* blog a University of Manchester researcher, Ian Plewis, as having demonstrated that there was no "spate of suicides" among Indian farmers.[22] Suicide rates, Plewis found, were no higher in India than in France or Scotland; farmers in the hardest-hit regions committed suicide at slightly lower rates than nonfarmers, and rates had remained stable after 2002, when the planting of GM cotton began. Quite apart from industry's ongoing claims about the safety of GM crops, this argument, if left unchallenged, would have turned the problem of farmer suicide into a nonproblem.

But things are rarely so simple at the nexus of wide-ranging socioeconomic and technological change. In April 2014, a joint University of Cambridge and University College London research study "found significant causal links showing that the huge variation in suicide rates between Indian states can largely be accounted for by suicides among farmers and agricultural workers."[23] That study did not specifically point to GM cotton as the culprit, but it did note that suicide was more prevalent in states with the most vulnerable farming populations, those working with plots of less than a hectare, under debt-ridden conditions, and growing cash crops such as cotton and coffee that are exceedingly sensitive to wide global price fluctuations.

While sociologists debated the causes and implications of distress on Indian farms, it fell to the nation's Supreme Court to address deficits in the regulatory structure for evaluating and controlling GM crops. Anti-GM activists turned to the court by means of a procedure known as a writ petition, asking it to block new field releases until a rigorous and transparent approval process was in place. In August 2013, a court-appointed technical panel recommended an indefinite moratorium on GM crops until several preconditions for safety were met. The panel report noted that India was not suffering shortages that warranted the introduction of GM food crops such as Golden Rice or Bt brinjal (eggplant). With regard to rice in particular, the panel cautioned that GM varieties might threaten biodiversity and endanger India's position as rice exporter, since there were as yet no provisions for segregating and labeling GM and non-GM varieties as desired by consumers in importing countries. The panel's report encouraged GM opponents but as just one more expression of expert opinion, it did little to resolve the underlying problems of politics and governance.

A DEFICIT OF TRUST

By the beginning of the new millennium, it was clear that GM agriculture faced tremendous hurdles because of a massive breakdown in trust between Western, especially American, agricultural scientists and industry and farmers and consumers across much of the rest of the world. The long-running controversy over genetically modified Golden Rice is symptomatic. The project to develop this form of GM rice originated in a humanitarian rather than a commercial impulse. It constituted one of the first ventures by agricultural scientists to engineer a plant to serve medicinal as well as nutritional needs. Both factors should have immunized Golden Rice against the sorts of debates that engulfed Monsanto's products, but they proved inadequate.

In the early 1990s, two eminent German-trained plant scientists, Ingo Potrykus and Peter Beyer, teamed up to create a variety of rice that could compensate for vitamin A deficiency, a condition that afflicts hundreds of thousands of children in Asia and Africa. Severe vitamin A deficiency can cause blindness and death, but it is easily preventable with proper nutrition in a child's early years. In wealthy countries, vitamin deficiency diseases of this kind are practically unknown, but poor children in regions subsisting on a largely rice-based diet with few supplementing vegetables are at much greater risk.

Potrykus and Beyer saw a golden opportunity to advance the science of plant genetic engineering, enhance the nutritional value of a major staple crop, and ameliorate a serious public health problem all at once. With support from the European Union, the Rockefeller Foundation, and eventually the Swiss-based biotech giant Syngenta, the scientists successfully modified rice to accumulate beta-carotene, a precursor to vita-min A. The beta-carotene gives the rice its pale golden color and its name. Eating this enriched variety is seen by propo-nents as a promising alternative to ingesting vitamin A in pills or other forms. Indeed, participants in the Golden Rice project point to failures in national public health initiatives as evidence that delivering vitamin A through local diets will work better than attempts to cure the problem through unreliable health care systems.

Nonetheless, transnational activist groups such as Green-peace vehemently oppose the introduction of Golden Rice on political, economic, and environmental grounds. Angry farm-ers uprooted plants and prevented field trials from being con-ducted in the Philippines, persuaded by Greenpeace campaigns that the GM crop poses a threat to local livelihoods. Scientists and mainstream media cried foul and deplored the acts of van-dalism, but it is not only extremists who point to unanswered questions. The Indian Supreme Court's technical panel on GM crop trials expressed a more nuanced judgment that scientific assessments of safety are not equivalent to—and cannot substi-tute for—the full-blown discussion of cultural values and socio-economic impacts that should have taken place in the early years of Golden Rice's development.

CONCLUSION

Few people would deny that understanding the genetic prop-erties of plants and animals has opened up a potentially revo-lutionary chapter in human interactions with nature, one that could lead to more sustainable ways of meeting the food needs of a growing global population. Nonetheless, the historically dominant approach to innovation in agricultural technology proved inadequate for addressing the range of issues laid bare by the GM revolution. Agricultural biotechnology as it has played out in recent decades raises critical questions about the concentration of power in select institutions and geographical areas through monopolistic control of production and through claims of intellectual property rights that are addressed more fully in chapter 7.

The conventional linear model of introducing technological change assumes that hazards can be more or less certainly identified in advance, that reliable methods exist for their assessment, and that the rising tide of innovation lifts all boats. Innovation, in short, is presumed to be a good in itself, and it is further presumed that the coupled realities of science and economics will keep industry from developing products that people find threatening or unpalatable. The emerging world of GM agriculture upended most of those expectations. The seeming precision of genetic engineering was not a good predictor of the complex interactions that became apparent during the production and commercialization of new crops and animals. Biology remains a great unknown, both at the level of individual organisms (as in the case of the Flavr-Savr tomato or the Beltsville pigs) and at the level of whole ecosystems. Repeated escapes and accidents, and the rise of biological resistance, have shown that presumptions about the controllability of plant genetic engineering were more optimistic than molecular biologists had anticipated at the Asilomar meeting. In hindsight, the earliest forays into GM agriculture look more like haphazard, even childish, tinkering with nature than like responsible innovation.

More seriously, these histories call attention to the extreme imbalance of power between those who design and market new products and those whose livelihoods are most vulnerable to technological change. Products such as GM corn that aided large U.S. biotech firms and big agribusiness failed to satisfy small-scale European farmers and consumers concerned about loss of biodiversity and threats to nontarget species. In India, the costs of the new technological system appear unsustainable for the small farmers who make up the backbone of the nation's agricultural economy, whose needs were not considered when companies such as Monsanto and Syngenta decided which pest-resistant or herbicide-tolerant crops would generate the largest share of profits on the Indian subcontinent. Even "humanitarian" products such as Golden Rice have not calmed the suspicion that such crops are in effect Trojan horses, through which hidden corporate interests will eventually steal in and seize control of a nation's entire grain production system.

In order for technological innovation to benefit the world's poorest and least empowered citizens, parallel innovations are needed in the theories and practices of global governance. The protracted controversy over GMOs between the United States and EU member countries suggests that the current global trading regime is not well equipped to address latent value conflicts surrounding innovation, let alone to overcome the burdens of economic and technological inequality. We will return to these points in the concluding chapter.

Chapter 5

TINKERING WITH HUMANS

A CODE REVEALED

In the middle of the twentieth century, life took on new meaning in science and society. Biology's most far-reaching transformation in centuries began in the heart of the proverbial ivory tower, in the Cavendish Laboratory, a famed research center for physics and biology at the University of Cambridge. There the British molecular biologist and biophysicist Francis Crick, then thirty-six years old, and his precocious junior collaborator from the United States, the twenty-five-year-old James D. Watson, solved a puzzle that had perplexed some of the brightest minds on both sides of the Atlantic for decades. Together, they worked out the structure of deoxyribonucleic acid (DNA), the basic matter that controls the development and functioning of all living organisms. DNA, they discovered, is a double helix with strands built of four bases in repeated fixed pairs. A two-page article reporting their landmark achievement appeared in the British scientific journal *Nature* on April 25, 1953. More than sixty years later, the ethical, legal, and social ramifications of that fundamental discovery are still being debated wherever law and policy intersect with the biological sciences and technologies—especially when the matters at stake concern human nature and human dignity.

Today the outlines of the great Watson-Crick discovery are familiar from countless works in the scientific literature, the news media, and recently also fiction, film, and television. The double helix of the DNA molecule is one of the best-known scientific images. It occupies an instantly referenced, bookmarked place in our repertoire of signs and logos, just as the word "life" does in our verbal lexicon. Like the images of planet Earth, the double helix needs no caption to be understood. Few now recall that eminent scientists such as Linus Pauling, chemist and double Nobel laureate, had imagined a triple helix structure for DNA before Watson and Crick announced their radical revision. Today even people with only the most rudimentary education in the life sciences are familiar with the basic mechanisms by which DNA functions as a motor of life and heredity.

Three features of the DNA molecule proved to be transformative for modern biomedicine and biotechnology. First, DNA is a container of information, often referred to as the code of life. That code is secreted in the genome, the full informational blueprint particular to an organism that is found in each of its cells. The best known and most widely studied components of the genome are the sequences of DNA called genes that provide the code for producing proteins; these in turn define the structure and functioning of the particular organism. The protein-coding genes, however, constitute only a tiny fraction of the total number of base pairs making up the genome—about 1.5 percent of the three billion or so in a human sperm or egg cell. The function of the noncoding parts of the genome remains as yet an open frontier for scientific investigation. Differences between species relate in important ways to the presence or absence of particular genes in the genome; the closer one species is to another, the larger the number of shared genes between them. Thus, chimpanzees and bonobos share about 99 percent of their genes in common with humans.

Second, the elegantly simple structure of DNA provides a recipe for replication. The DNA molecule is built of four bases called adenine, thymine, guanine, and cytosine, abbreviated for convenience as A, T, G, and C. These bases bond together in fixed pairs, like ladder rungs along the twisted double helix, A with T and G with C. Placed in the right chemical environment, the helix unwinds like a long zipper being unzipped. The resulting single strands of DNA can then recombine in the same sequence, each base retying itself to its chemical opposite number to create two identical helical structures where previously there was just one. In this way, the strand reproduces itself exactly, a fact that Watson and Crick noted with epic understatement: "It has not escaped our attention that the specific pairing we have postulated immediately suggests a possible copying mechanism for the genetic material."[1]

Third, important physical properties and propensities of human beings become more transparent when represented in terms of the genetic code. The code offers in effect a different language in which to express certain aspects of human identity and human nature. Is someone a carrier, for example, of the gene for Huntington's disease or a gene that predisposes one to a higher risk of breast cancer or early-onset Alzheimer's? Knowledge of DNA also makes bodies more manipulable. The structure's extreme simplicity makes certain kinds of testing and rebuilding feasible. It is possible in principle to cut and paste together bits of DNA from diverse sources without impeding the ability of the entire strand to replicate itself or to carry out its information transfer functions. If the inserted DNA is a gene with coding properties of its own, those properties are simply transferred into the reengineered host DNA. For example, bacterial DNA that has pieces of human genetic material spliced into it will produce human proteins, as instructed by the transferred component, along with all of the bacteria's own natural proteins. In principle, human DNA can also be modified in this way to produce new traits. If the modification is done in a sperm or egg cell, the inserted DNA and the traits it codes for will be passed on to any subsequent generations arising from the manipulated cell. This is called germline genetic engineering, and it is currently prohibited for humans in most parts of the world that have thriving research capabilities in biomedicine.[2]

No scientific discovery, however revolutionary, moves into applications through ideas alone. Many further inputs are needed to achieve that translation, from new skills and techniques to material resources, capital, and institutional support. A major breakthrough at Stanford University in the early 1970s facilitated the transfer of knowledge of what DNA is and how it physically functions into commercial products. This was the technique of recombinant DNA (rDNA), popularly known as "gene splicing," which removes DNA sequences, usually genes, from a source organism and implants them in a host organism. Inside the host, the foreign DNA replicates along with the organism's own. Developed by Stanley Cohen and Paul Berg of Stanford and Herbert Boyer of the University of California at San Francisco, the technique of gene splicing enables scientists to transfer DNA between species, creating organisms that could not have existed in nature, such as bacteria that produce human insulin, or plants with genes from phosphorescent jellyfish that glow in the dark. Stanford University filed for a patent on the rDNA technique in 1974, naming Cohen and Boyer as the inventors.[3] Awarded six years later in 1980, the patent earned the university some $255 million before expiring in 1997. In that time, the use of rDNA molecules permeated the pharmaceutical and agricultural industries, revolutionizing both.

Manipulations of human biology raise even more complex issues for ethics, law, and policy than those involving plants, seen in chapter 4. In biomedicine, the greatest fear centers on violating human integrity and eroding the fundamental meaning of being human. Knowing human bodies in a new way, through a person's genetic code, opens up the prospect of unprecedented intrusions on cherished rights of liberty, equality, and privacy. At the same time, these rights themselves get redefined in the light of scientific and technological advances.[4] Medicine's primary mission is to make sick people whole and at-risk people stay healthy, but the line between using gene therapy to cure defects and genetic enhancement to endow people with superhuman capabilities is blurred and controversial. Reproductive medicine assists infertile people to conceive, thereby fulfilling what is for many a lifelong dream; but it also opens the way to weakening the meaning of family relationships and diluting the responsibilities of parenthood. The rise of bioethics as a discipline in the late twentieth century testifies to the seriousness of these concerns. Professional ethicists have gained increasing authority to develop guidelines and set limits on how far society should go in tinkering with human biology, but many questions remain disputed or unsettled, including the fundamental question of how much authority democratic societies should delegate to experts in ethics.

TRANSPARENT BODIES

As in any revolution, the great promises of post-genomic biomedicine trailed shadows in their wake. Worries in this case centered not only on careless and irresponsible industrial behavior, driven purely by short-term profit interests, but on an imaginary of science out of control and an alliance between technology and governmental or corporate power that could threaten the very notion of being human. Would curiosity and sheer delight in tinkering with nature tempt scientists to stray into experimental zones abhorred by society, and were new institutions needed to prevent such overreaching? Would increasing genetic knowledge exacerbate already troubling imbalances between individuals and institutions? Would it threaten rights to liberty and privacy? Would it further stigmatize marginal groups, or would it give them resources to become more self-governing? To evaluate the emerging answers to these questions, we must look first at the devices with which genetics and biotechnology have made human bodies more readable and, potentially, also more open to technological intervention.

The four-letter alphabet of the genetic code offers biomedicine a powerful tool for locating some of the precise sources of hereditary illnesses. Genes code and transmit information in ways that help explain variations not only among species but also between individuals of the same species. Genes for a given trait may occur in several forms, each associated with an observable difference, for example, in eye or skin color among humans, or in the wrinkled versus round seed shapes that Gregor Mendel, the pioneering nineteenth-century genetic scientist, observed in pea plants in his garden in the Czech city of Brno. These variants are known as alleles. Many genetic diseases have been linked to alleles that cause malfunctions in information transfer. Those defects impair an organism's ability to produce proteins that are essential for bodily growth and health or to fight the onset of cancer and other illnesses. For example, women of Ashkenazi Jewish background who have inherited harmful versions of the so-called BRCA1 and BRCA2 genes have a markedly higher risk of contracting breast and ovarian cancer because the mutant genes do not produce the proteins that help suppress those tumors. Finding and, if possible, correcting these errors in information transfer are the central goals of genetic medicine.

The study of genetic variation gained momentum with the completion of the Human Genome Project, an international effort to map and sequence the entire set of base pairs that make up human DNA. In the United States, the National Institutes of Health (NIH) and the Department of Energy funded the project at the federal level, with James Watson serving as the NIH program's first director. Officially launched in 1990, the project produced a draft sequence in 2000, well ahead of the fifteen years that scientists had originally estimated for completion. Unexpectedly, what began as public "big science" turned into a race with the private sector. J. Craig Venter, a onetime NIH employee, broke away and founded his own company, Celera Corporation, to prove that sequencing could be done faster and cheaper by using a different technological approach from the government's.* The NIH responded with its own improvements, further expediting the process, and Venter established himself through this maneuver as one of biotech's most famous entrepreneurs. More recently the compilation of genetic material in large

---

*The NIH used a "hierarchical shotgun strategy," which involved first locating large fragments of DNA within the genome and then shredding and sequencing them. By contrast, Celera used the "whole genome shotgun strategy," which involved shredding the entire genome, sequencing pieces with a massive array of computing power, and eventually piecing back the results. Originally in fierce competition, the two methods were eventually used to complement each other.

collections known as biobanks has bumped up the study of genetic variation from the scale of individuals and families to populations with specific genetic traits, as well as groups sharing broader ethnic and racial affiliations.

Tests and Privacy

The first doors that genetic medicine unlocked were diagnostic. Knowing that specific genes are associated with particular diseases or predispositions to disease can help people make earlier, more informed medical choices. Should a woman who carries the mutant breast cancer genes undergo prophylactic mastectomies to reduce her risk of cancer? Women may choose this drastic measure to be freed from anxiety, especially if they have previously lost a mother or sister to those unforgiving illnesses. The actress and global celebrity Angelina Jolie opted for a well-publicized double mastectomy in 2013 when she discovered that, as a bearer of the BRCA1 gene, she had an 87 percent risk of developing breast cancer and a 50 percent risk of developing ovarian cancer.[5] Her story no doubt induced others to take genetic testing more seriously, just as U.S. First Lady Betty Ford's public discussion of her own struggle with breast cancer had brought the disease into the open in the mid-1970s.

The consequences of having the mutant BRCA genes are relatively well understood, but not all genetic knowledge is as unambiguous. Genetic test results can be disturbingly hard to interpret even if they are responsibly acquired and communicated. Patients may despair or make unwise choices if genetic tests reveal markers of diseases for which modern biomedicine as yet holds no cures, such as Huntington's disease or Alzheimer's. Persons who bear such predisposing conditions sometimes prefer ignorance, believing that knowledge can only bring stigmatization or loss of hope; indeed, some have argued that people have a right *not* to know of presymptomatic conditions that cannot be altered or relieved. Further, since some diseases disproportionately afflict specific racial groups, such as sickle-cell anemia among African Americans, the fear that genetic knowledge will exacerbate latent racism has accompanied genetic medicine from its earliest years.[6]

The ethical quandaries that arise in connection with genetic diagnosis were traditionally addressed within the framework of the doctor-patient relationship, which delegates to physicians and medical institutions the duty to care for patients' well-being. Hospitals in industrial nations now routinely provide genetic counseling services to ensure that patients who receive diagnoses will properly interpret the results and fully understand their treatment options. Since test results convey at best a probability of harm, and rarely absolute certainty, counseling aims to help patients reach decisions that are neither too risk averse nor too complacent.

A sharp drop in the cost of genetic tests, however, has allowed control of patients' genetic information to slip out of the health care system into normal commerce. Beginning around 2007, private companies moved to capitalize on what they saw as a market potential in direct-to-consumer (DTC) testing. Companies such as 23andMe and deCODE began offering consumers the opportunity to send away saliva samples and receive back information on either their heredity or their vulnerability to specific genetic diseases. The DTC testing market is expected to grow substantially in coming years, although much uncertainty remains about its reliability, health benefits, and legal status. Worries for consumers include questions about the accuracy and clinical validity of the tests, the privacy of the information held by companies, and the fate of the information if a DTC company goes out of business or is sold. In 2013, the Food and Drug Administration (FDA) moved somewhat belatedly to assert regulatory oversight, declaring that DTC tests would be regarded as medical devices requiring prior approval. Months later, FDA ordered 23andMe to stop marketing its tests until approval was obtained.

Privacy concerns are not limited to the context of DTC testing. From the early days of gene testing, people feared that their genetic information might fall into wrong hands, allowing employers and others in powerful positions to discriminate against them. Knowing that a prospective employee has a heightened risk for a potentially costly medical condition could lead to denial of insurance or loss of a job. Yet, as long as genetic privacy remained a question of civil liberties alone, lawmakers seemed reluctant to move on the issue. The ground shifted with the conclusion of the Human Genome Project. Brighter prospects for precision medicine, or treatment based on each individual's personal genetic profile, persuaded U.S. legislators that stronger privacy protections would encourage more widespread use of genetic testing and lead to profitable advances in clinical and pharmaceutical research. There was, in short, a growing market demand for privacy as an inducement to persons reluctant to be tested without adequate safeguards. The Genetic Information Non-discrimination Act (GINA) of 2008[7] offers federal protection against misuse of genetic information by employers and insurers, though the law excludes life, disability, and long-term care insurance—significant exclusions for an aging population. GINA likewise does not protect information acquired by DTC testing companies from people making voluntary use of their services.

Populations as Data

If the conversion of physical traits into the abstract language of "information" creates problems for governance, added puzzles lie in store as genetic information becomes intertwined with the data revolution. Biobanks, or large-scale collections of biological samples and the information they contain, pose a host of new questions for which societies as yet have no definite answers. The concept of banking human tissues predates the genetic age. Blood and sperm, for example, were collected and stored in so-called banks, available at need to persons requiring transfusions or male infertility treatment. Genetic biobanks differ from these precursors in that they are storehouses of information that is more or less permanent, as well as for physical materials that may decay. The commercial value of their holdings derives in large part from the fact that the information can be mined for diagnoses and uses unconnected with any individual donor or beneficiary.

Modern biobanks raise issues of ownership, consent, and privacy that were not relevant to the management of blood and sperm banks of an earlier era. One of the first attempts to build a genetic biobank occurred in Iceland, but the plan never took off as conceived and its failure remains instructive.[8] Three factors made the prospects for creating such a resource in Iceland seem rosy at first. The country has, to begin with, a small and relatively self-contained population (just under 300,000 at the time of the proposal) and a long tradition of maintaining genealogical records. Familial traits and population-wide variations could therefore be identified with relative ease. Moreover, ancestry data could be set beside individual health records in a country with a well-functioning medical system. Iceland also had the infrastructure to enable genetic data to be collected from every citizen, providing the third crucial element for the proposed national Health Sector Database (HSD).

Spearheaded by deCODE Genetics, a biopharmaceutical company headed by the charismatic scientist-entrepreneur Kári Stefánsson, the HSD was conceived as a public-private partnership that won enthusiastic support from a majority of Iceland's Parliament. Under the 1996 Health Sector Database Act, the public sector was to contribute data and the private sector, effectively deCODE, was to provide the know-how to convert information into lifesaving drugs against diseases such as multiple sclerosis. Since enrolling virtually all of the nation's population was key to the success of the database, the designers chose an opt-out system: each citizen's medical and genealogical information would be included under a principle of "presumed consent," unless he or she explicitly chose to opt out at the time of initial enrollment.

About 10 percent of Icelanders immediately elected not to participate, but not content with merely opting out, some pursued legal action to stop the project. Plaintiffs objected to being enrolled as unconsenting research subjects and to the lack of provisions for withdrawing once in. By contrast, widely used ethical guidelines allow patients enrolled in clinical trials to withdraw at will. Litigants expressed particular displeasure at the inclusion of records from deceased persons who could never have given consent, and whose information would inevitably compromise the privacy of family members sharing the same genetic heritage, even if those family members themselves withdrew from the database. Complainants also faulted deCODE's monopolistic control over an entire nation's medical and genetic records, as the sole licensee for a twelve-year period. In 2003, the Icelandic Supreme Court struck down the HSD Act. Its provisions were never fully implemented, although deCODE continued to do research on genetic variation, producing significant publications on the genetic causes of schizophrenia and cardiovascular disease.

Despite this setback, biobanks have continued to grow in popularity in other countries while also presenting ongoing ethical challenges. One of the largest and most ambitious such initiatives, UK Biobank, began in 2006 as a public-private partnership between the UK government and the Wellcome Trust, the nation's largest private funder of biomedical research. Critics worried about the overhyping of genetic information as the primary instrument for diagnosing critical illnesses, a worry fed in part by mounting knowledge that genes alone may be less determinative of health outcomes than was thought before the sequencing of the human genome. Questions arose about the breadth of the consent demanded of participants and the eventual opening of the biobank to commercial researchers and possibly law enforcement. Nonetheless, by 2009, UK Biobank had succeeded in enrolling its target population of half a million subjects between the ages of forty and sixty-nine, establishing the UK as the world's leading nation in the number and size of its biobanks.

U.S. initiatives thus far remain more local, reflecting the absence of a national health care system with resources and incentives to centralize the collection of genetic information. Still, such data gathering has powerful support among one of the world's leading biomedical research communities. In 2009, the prominent medical scientist and bioethicist Ezekiel Emanuel, who also served as health policy adviser to President Barack Obama, coauthored an article arguing that citizens have an affirmative duty to participate in clinical research.[9] Since we are all present and future beneficiaries of increased medical knowledge, Emanuel and his coauthors maintained, such knowledge is a public good, and we should all in effect "pay" with our bodies for its production, much as everyone has a duty to serve as juror in the service of law enforcement. On this public good theory, it would be relatively straightforward to require everyone in a country to provide genetic information to a national biobank.

Emanuel's vision remains for now in the realm of imagined futures, one that some might find too close for comfort to that of *Brave New World*, Aldous Huxley's dystopic novel about a state-managed biological citizenry. The proliferation of biobanks created and managed by private and nonprofit organizations, however, is already raising vexed questions of ownership and control. Who owns life? And precisely what rights does the concept of ownership entail when claims encompass physical material from living organisms, as well as information extracted from them? These questions are addressed in greater detail in chapter 7, which deals more generally with intellectual property rights in a high-tech age.

ASSISTED REPRODUCTION

While molecular biologists such as Cohen and Boyer were tinkering with the structure of DNA, thus laying the foundations for the modern biotech industry, a pair of British scientists were experimenting with one of life's most basic and intimate processes: human reproduction. Robert Edwards, a developmental biologist at the University of Cambridge, teamed up with Patrick Steptoe, a gynecologist at Royal Oldham Hospital in Manchester, to create a novel technique for treating human infertility, known today as in vitro fertilization (IVF). Their pathbreaking work involved in effect removing the initiating segment of the natural human reproductive cycle, conception, from inside a woman's womb to the artificial setting of glass vessels in a laboratory. This is why Louise Brown, born through IVF at Oldham on April 25, 1978, was instantly dubbed the world's first "test tube baby."

For many, Louise's birth was a miracle promising biological parenthood to couples otherwise incapable of having children of their own. Indeed, thirty years later, worldwide estimates of the number of babies born through technologically assisted means had reached five million. Clearly a global demand exists, at least among people prepared to pay the hefty price of IVF treatments. Yet, from the earliest years, thoughtful observers wondered about the wider implications of transporting one of the most natural functions of the human body into the domain of manipulation and experiment. How would this move alter the meaning of being a parent or being a child? Would childlessness come to be stigmatized in a society where "birth control" means not only preventing the births of undesired children but also having children only within bounds that society deems to be natural, such as within heterosexual marriages involving couples of normal childbearing age? What additional pitfalls might lie along the slippery slope of converting conception, gestation, and childbirth into processes that humans can command and control?

One line of experimentation occurred shortly after, though not as a direct consequence of, the first barrier-breaking success of IVF. This was the practice of surrogate mothering, or surrogacy, which brought a consenting third party into a relationship with the express purpose of bearing a child for another couple. The practice won widespread notoriety in the 1986 U.S. case of *Baby M*, so named for Melissa, the baby at the heart of the controversy. Mary Beth Whitehead, a married mother of two living in New Jersey, agreed to conceive and carry a child for William and Elizabeth Stern. The Sterns, a highly educated professional couple, entered into the arrangement because they feared that Elizabeth's multiple sclerosis would not allow her to carry a child safely to term. Whitehead, however, felt unable to give up the baby after she was born, setting the stage for a bitter legal struggle that went up to the Supreme Court of New Jersey. In a first of its kind ruling, the court held that surrogacy contracts were invalid under New Jersey law, but it nonetheless awarded custody to the Sterns on the ground that this would serve the best interests of the child. Although Whitehead retained visitation rights as the child's biological mother, their relationship became increasingly distant. At the age of eighteen, Melissa Stern chose to be adopted by Elizabeth and thus legally extinguished her ties to Whitehead.

The ethical questions first raised by Baby M intensified as surrogacy, together with IVF, allowed many previously unthinkable parent-child constellations to be formed, for example, with postmenopausal mothers or gay and lesbian parents or women in other countries. Twenty years after the introduction of IVF, a technological development involving reproduction in nonhuman mammals introduced yet more quandaries about ethical limits on the manipulation of human life. This was the birth of a sheep named Dolly in Edinburgh's Roslin Institute, a leading British center for research on animal health and welfare and on the implications of that work for human health. Unlike normal mammals, Dolly was an exact genetic replica, or clone, of her mother. Plant breeders have for centuries propagated specimens with highly desirable traits from cuttings that preserve the genetic identity of the original. Sexual reproduction in mammals, however, combines genetic material from the mother's egg and the father's sperm to produce an offspring half of whose genes come from one parent and half from the other. Exact copying was not possible through traditional animal husbandry, even though many had wished to replicate the characteristics of an especially fleet race horse, a cow or pig with exceptionally lean meat, or a beloved pet animal. That barrier fell when a team of scientists led by Roslin's Ian Wilmut succeeded in cloning a newborn lamb from an adult sheep's cells, transferring intact to the child the mother's entire genome.

The process of cloning, in animal or human cells, builds on techniques of IVF that are by now well established, but with a new twist. First, the nucleus of a cell from the body of an adult, containing all of that animal's genetic code, is removed and placed inside the egg of another member of the same species. The modified egg is reimplanted in the new mother's uterus, where it comes to term and is born not as a genetic offspring of the mother who gestated it but as a younger replica of the original adult from whom the egg's nucleus was taken. The technique's commercial potential was immediately apparent to farmers—it would remove the element of chance from breeding genetically desirable animals—but so were some more disturbing implications. If cloning was possible in mammals like sheep, it probably could also be done with humans. A scenario once contemplated only in science fiction, most notably in Aldous Huxley's novel of graded, industrially produced humans, now seemed well within the horizon of the possible.

Concerns about breaking this barrier in reproductive biology rose to the highest levels of political salience. In the United States, President Bill Clinton quickly asked his ethics advisory panel to analyze the implications of human cloning.[10] His advisers recommended a ban on cloning human beings that has since been endorsed by, among others, President Barack Obama and remains intact in most advanced industrial countries. Other closely related techniques, however, proved more debatable, revealing substantial differences of opinion about the treatment of the developing human being, from embryo to fetus to eventual member of a family and a nation.

One such issue concerns the creation of the "three-parent embryo." The purpose of this technique is to produce a healthy embryo that will not carry genetic diseases transmitted by mitochondrial DNA (DNA lodged outside the nucleus) from the mother's egg cells. To achieve this result, the nucleus from the future mother's egg is extracted and placed inside another woman's donated egg cell from which the nucleus has been removed. The embryo and future child resulting from this fusion of two human eggs contain the intending mother's genome but without the risk of inherited mitochondrial disease from the rest of the cell. The technique was approved for use by the UK Parliament in February 2015 following wide public consultation, but it remains controversial in the United States and forbidden in Germany.

Another live debate concerns a possible rollback of explicit or tacit prohibitions against germline gene editing in most biomedically advanced democracies. Here, the destabilizing impetus comes from CRISPR (clustered, regularly interspaced, short palindromic repeat) technology, a technique that allows precisely targeted and rapid modification of genes to remove diseased segments and replace them with healthy alleles. If used to repair diseases in human IVF embryos, the resulting altered trait would be passed on by any child to its descendants, violating prohibitions on such modifications in legally binding documents like the Council of Europe's Convention on Human Rights and Biomedicine. What kind of body and which forms of deliberation are best suited to setting limits on the use of such a tempting and widely applicable technology? The question calls for new thinking about the right forms of governance as well as the proper domain of expert prediction. To date, the discussion has been led more by the elites of global science than by law, religion, or ethics.[11] The global public whose future genetic heritage is under discussion remains formless, unrepresented, and unable to speak.

DESIGNED AND SELECTED LIVES

Like the discovery of rDNA, IVF and cloning opened the way to cascading discoveries and inventions with far wider social implications than treating one couple's infertility or breeding one particularly promising strain of farm animal. Labs and clinics were now in a position to intervene in a dimension of the human reproductive cycle that would-be parents once regarded as their private domain. The embryo detached from the mother was itself an object of uncertain status, of great moral significance as a potential human but lacking the capacity to think or feel and unable to survive without technological supports. How far should researchers go in studying these entities, especially when their work would destroy human embryos? What principles and limits should govern the choices made by IVF users? Who would determine what should be done with nonimplanted embryos? Three areas of conflict help illustrate the evolution of the questions and the fluidity of the answers and approaches that have emerged across advanced industrial nations that ostensibly subscribe to the same basic values. These are the issues of research with human embryonic stem cells, designer children, and appropriate parenthood.

Human Embryonic Stem Cells

The IVF process, as it has been standardized in medical practice, produces more embryos than should be, or even can be, implanted in the mother's womb in order to maximize the chances of a safe pregnancy and healthy birth. The remaining "spare" embryos can be used for other purposes, for other infertile couples wishing to carry them to term or for the derivation of human embryonic stem cells (hESCs) that could be used in treating previously incurable diseases.

The sudden availability of human embryos as organisms detached from the reproductive cycle opened up a whole new frontier in biomedical research. Stem cells are early-stage cells taken from an organism before it has begun the complex process of differentiation that leads to the mature organism whose properties are encoded in the genome. In particular, cells derived from a human embryo while it is still in the blastocyst (eight-cell) state are not yet specialized to become blood, brain, skin, or bone. They are "pluripotent," that is, capable of taking virtually any form depending on the context into which they are implanted. A dawning hope in biomedicine is that these highly adaptable cells might be used one day to cure many kinds of human diseases, such as Parkinson's, Alzheimer's, leukemia, or infant diabetes, caused by the body's inability to generate healthy new cells. That hope moved a giant step closer to reality when James Thomson of the University of Wisconsin isolated the first hESCs in the early 1990s. Since then centers for stem cell research have sprung up all over the United States and many other leading industrial nations, accompanied in most cases by some degree of scandal and controversy.

The core ethical problems in human embryonic stem cell research have to do with the status of the embryo from which an hESC line is derived, the consent of the persons from whose germ cells the embryo was produced, and the potential misuse of stem cells for research or therapies that a majority of people find objectionable on moral grounds. Approaches to resolving these issues vary dramatically across countries, as do the resolutions reached by ethics bodies and lawmakers. These divergences in turn reflect widely different understandings of the risks of using embryos as research subjects, as well as the degree of autonomy society is willing to grant to science in choosing directions for research and technological development.

Stem cell derivation ordinarily destroys the embryo from which therapeutically active cells are extracted. For those who see the embryo as a form of already-human life, this process therefore amounts to murder. In a formulation often used by former U.S. president George W. Bush, it means destroying life to save life. Three very different policy responses in the United States, Britain, and Germany illustrate how even the constitutional democracies of the West stand far apart with respect to these issues. U.S. courts and legislatures have taken no firm position applicable nationwide on the question of when human life begins. U.S. institutional mechanisms for making and implementing research policy are highly decentralized and not governed by any single legislative mandate. Yet, in a spillover from the acrimonious, long-running abortion debate, a little publicized federal law known as the Dickey-Wicker amendment prohibits federal funding for the creation of embryos for research and for research that destroys or discards human embryos. This amendment was attached to an NIH appropriations bill in 1996 and has been renewed each year since. Consequently, in the United States hESCs can be created for research only with funding from private or nonfederal sources. With some twists and turns over the years, the NIH has continually funded research on approved stem cell lines that were developed in accordance with federal law. NIH funds may be used only by institutions that maintain an adequate Embryonic Stem Cell Research Oversight (ESCRO) structure to review research protocols and approve them in accordance with applicable state and national guidelines.

Britain, by contrast, formally adopted the position that the embryo before it is fourteen days old has a moral status different from its status after that cutoff point. The early embryo, UK policy holds, does not have characteristics that should trigger concerns about protecting human life as such. After fourteen days, however, the embryo begins to develop a primitive nervous system, and it can no longer be regarded as an object that is clearly not a human being. From that point on, it is at least a prehuman. This "fourteen-day rule" may well be the most important decision ever made by a bioethics body, a committee chaired by (and named for) the Oxford moral philosopher Mary Warnock.[12] The Warnock Committee recommended that Britain create a new body with the duty to supervise all activities related to embryos. A 1990 law established the Human Fertilisation and Embryology Authority (HFEA), which has for a quarter century licensed all IVF clinics, reviewed all applications for embryo research, and regularly consulted with the public so as to form a consensus on how to press forward on the controversial frontiers of human reproductive research.

Germany, too, defines the beginning of human life in biological terms, but its definition is markedly different from Britain's, and it satisfies a demand for precision in the Basic Law that underwrites the post-World War II German state. Article 1.1 of the Basic Law is Germany's constitutional response to the Holocaust: it states in categorical terms that human dignity is inviolable and that the state has a duty to protect it. To meet this prime mandate, German lawmakers and courts eventually decided that human life begins at the precise moment when the nucleus of the egg fuses with the nucleus of the sperm, since that is when a genetically distinct entity comes into being. It follows that deriving stem cells from human embryos would amount to destroying potential human lives; hence such derivation is forbidden by law. Importation of hESCs, however, is treated as a different matter. German researchers may work with imported hESC lines created abroad, since this removes from German science the taint of destroying the embryos from which those cells were derived. The result is an uneasy division of moral responsibility, similar in effect to that in the United States—except that German law banishes the destruction of embryos firmly outside the nation's borders, whereas U.S. law absolves the federal government of responsibility for supporting embryo death while allowing states and private sources to continue funding such research. Each country's ethical settlement reflects distinctive national expectations about the right way to balance the competing demands of science, religion, and public health.

Designer Babies

IVF and cloning, together with procedures for genetic testing, opened up the possibility not only of having a baby at all, even for infertile couples, but also of choosing, within limits, what kind of baby to have. Standard IVF procedures create more embryos than are implanted for reproductive purposes. This allows for a certain degree of conscious selection, to ensure that the baby ultimately carried to term will conform to parental needs and desires. Most important to parents is the desire for children who are not at risk for genetic diseases or disabilities that could be transmitted via a normal pregnancy. Fetal cells are now routinely subjected to a battery of tests to determine whether the fetus is carrying a gene for severe birth defects such as Down's syndrome or Tay-Sachs disease. Parents can choose to abort an abnormal fetus and hope for a disease-free baby the next time they conceive, but this can be stressful and heartrending for parents, as well as risky for the mother-to-be. The technique of preimplantation genetic diagnosis (PGD), used together with IVF, seeks to forestall the need for a potentially traumatic or dangerous abortion. PGD allows in vitro embryos to be screened for genetic abnormalities before they are implanted, so that parents can embark on a pregnancy with greater certainty that they will eventually have a healthy baby.

The opportunity for embryo selection gives rise to a number of choices and dilemmas. Embryos can in principle be screened for characteristics that bear on human diversity but have nothing to do with disease. The most notorious example is the use of prenatal or preimplantation screening to allow parents to preselect the sex of their offspring. Though many support sex selection for the purpose of achieving gender-balanced families, deep-seated cultural preferences for male children have led in some societies to a troubling population-wide overrepresentation of male in relation to female births—even when, as in India, fetal sex selection is prohibited by law. With further advances in technology, in vitro embryos could potentially be modified to introduce desired traits. As yet, the making of such "designer babies" remains largely hypothetical, but examples of what parents could do, assisted by sympathetic bioethicists and skillful clinicians, are already on the horizon.

PGD can be used not only to test fetuses for their own genetic abnormalities but to determine whether their tissues are compatible with those of an already extant child who needs a transplant from a healthy donor in order to live. The ethics of selecting for such "savior siblings" have spilled into popular culture, for example, in Jodi Picoult's best-selling novel *My Sister's Keeper* and Kazuo Ishiguro's wrenching *Never Let Me Go*.[13] Picoult depicted the relationship between one sister with acute leukemia and another conceived to be her blood, bone marrow, and eventually kidney donor. Ishiguro blended the classic British boarding-school novel with science fiction in a haunting story about a group of schoolchildren who live and die as organ donors for "normal" people. Both works express painful, as yet unresolved anxieties about a future in which people may be brought into existence mainly to serve others' medical needs—in Kantian terms, as means or instruments rather than as ends in themselves.

Real life fortunately operates at a slower pace than fiction, and a donor class of the kind Ishiguro imagined may never come into being. Individual cases, however, have already arisen. Britain in 2010 reported the first successful treatment of an incurable disease using one sibling's tissue to save another. Megan Mathews was born with a rare blood disorder that required her to receive transfusions every week; she was cured with a bone marrow transplant from her newborn brother Max. The Mathews parents conceived Max using IVF and PGD to determine that the implanted embryo was disease-free and would be a good tissue match for their ailing daughter. Years of legal wrangling preceding this medical success story exposed an irreconcilable gap between parents who believe they are creating a mutually supportive and loving family, in which the savior sibling is merely "helping" an older sister or brother, and those who, like Britain's pro-life activist Josephine Quintavalle, fundamentally oppose the creation of any child to serve a therapeutic purpose for another.

The Family Reconceived

The Baby M case brought to the front of popular consciousness the possibility of a three-parent family—a move that some thought gave women the same access to infertility treatment that infertile men had enjoyed for years through artificial insemination by donor sperm. But few could have predicted how surrogacy combined with IVF would redefine the constellation of possible families within decades, or that the legal and ethical ramifications of surrogacy would continue to reverberate undiminished into a new century.

Most of the disputes surrounding assisted reproduction center on the degree of a society's commitment to traditional notions of the family, such as the heterosexuality of the parents, their age, and their marital relationship. Disagreements also persist about whether these issues should be regulated by law or be left to private arrangements between clients and clinics. The table below displays how different combinations of these two variables can lead to substantial divergences in the kinds of families a society will tolerate within the spectrum of the "normal."

TABLE 2:

Norms / Form of Control	Traditional	Nontraditional
State Regulated	IVF allowed with varied conditions: • Only heterosexual couples • Marriage required • Women's age regulated • Limit on number of embryo implants Surrogacy banned or not recognized by law.	Permissive laws, but with some constraints, e.g., no commercial selling of eggs.
Unregulated (clinics set standards)	IVF voluntarily sought by traditional heterosexual couples.	Homosexual couples allowed. Marriage not required. Women's age not regulated. Number of implants not regulated.

Among Western nations, the United States has been among the most permissive and decentralized in regulating the new reproductive technologies. There is no federal law concerning the use of reproductive technologies at the national level, and state supervision is often lax. Perhaps predictably, therefore, the United States has emerged as the world's leading site for reproductive experiments, including the first cases of gestational surrogacy,[14] an arrangement in which the pregnant woman carries a child who is biologically unrelated to herself, as well as the first recorded live birth of IVF octuplets, and numerous births for women over fifty years of age.

Lax regulation not only encourages experiments with multiple forms of parent-child relationships but may also lead to difficult legal disputes when parties opt out of contractual agreements (as in the Baby M case) or to textbook cases of reproductive irresponsibility when couples go their separate ways after the birth of an IVF child. A particularly bizarre example was the story of Jaycee Buzzanca, a baby girl born in California in 1995, who spent three years legally "parentless," although she arguably had filial claims on five different adults. Jaycee was born of a donated embryo from an unknown couple to an unrelated married surrogate mother who had contracted to carry the child for another married couple, John and Luanne Buzzanca. The Buzzanca marriage dissolved before Jaycee's birth, and John disclaimed any responsibility toward the child. A trial court initially held that Luanne could not be the legal mother as she desired, because she had neither given birth to the child nor was genetically related to her. The Supreme Court of California ultimately clarified the law, ruling that the "intended parents" who initiated the medical procedure were also the child's legal parents regardless of any subsequent changes in their intentions.

Questions about parentage merged with questions of national affiliation and citizenship in a series of cases involving couples from strictly antisurrogacy nations, like France and Germany, who acquired children through surrogacy in countries with more permissive laws, like the United States and India. Two French families, the Mennessons and the Labassees, tested the unsettled jurisdictional waters when they chose to have children in California, taking advantage of that state's liberal environment for surrogacy. France initially denied citizenship to their daughters, who by birth were entitled to be U.S. citizens. In June 2014, the European Court of Human Rights (ECHR) ruled that France was obliged to grant citizenship to both families' children, because they would otherwise grow up with diminished

identities and rights vis-à-vis other children of French parents.
The ECHR saw this as a violation of article 8 of the European
Convention on Human Rights, granting children the "right to
respect for private and family life."[15] Critics complained that
the ECHR was undermining France's domestic public order
by signaling to parents that they would receive a free pass for
engaging in conduct abroad that was forbidden at home. Others
countered that the case should be narrowly construed as being
about the best interests of the children in question and that no
larger issues were at stake concerning French sovereignty or the
validity of the French civil code. The French government in any
case decided not to contest the ECHR decision with respect to
the Mennesson and Labassee daughters.

LIVES AND LAWS

A major theme of this chapter is that technology, once it escapes
from the closed worlds of labs and field tests, becomes to some
extent everyone's property. Widening uses of technology, moreover, change the users' sense of their own identity and potential
as they come to understand what they can do, and even who
they are, in novel and unpredictable ways. Having new tools
thus enables people to imagine and realize new futures. For the
most part, however, ethical and regulatory analysis still pro-
ceeds on a linear path, plugged in somewhere between invention and the market, oblivious to the feedbacks and creative
extensions that occur as productive inventions embed themselves into complex and dynamic social contexts.

The decades following the unraveling of the genetic code wit-
nessed intense and widespread social experimentation with the
stuff of life. Not only technical experts but ordinary members

of society showed that they were prepared to act upon genetic knowledge and know-how, creating new market demands, engaging in new partnerships between citizens and scientists, and adopting new practices for procreating and forming families. What emerged very clearly from this ferment is that disentangling the laws of life went hand in hand with creating new entanglements between lives and laws. Age-old questions resurfaced in novel forms (when does life begin or end in the stem cell era?), while questions that people had considered long settled were thrown into doubt (who is a natural mother?). These questions could not be answered by science alone. They belonged as much, if not more, to politics, ethics, and law.

It is appealing to see law and ethics in these cases as being in a constant race to keep up with science and technology. That conclusion seems obvious on its face: after all, there were no rules already in place to resolve the Stern-Whitehead controversy over Baby M or the puzzle over how to assign parental rights in the case of Jaycee Buzzanca. To see law and public morality as always lagging, however, leads us into the trap of technological determinism. It suggests that technology sets its own moral codes, and public values simply catch up later. What we have seen far more often in this chapter is that, confronted with novel ways of characterizing and manipulating the stuff of life, people have striven with energy and ingenuity to rearticulate their fundamental moral commitments: to the preservation and protection of life; to upholding human dignity; and to maintaining institutions such as motherhood and fatherhood, albeit with a wider range of possibilities and imaginations than before. In each case, these sociotechnical exercises in self-fashioning have drawn simultaneously, and in equal measure, on the material resources of technology and the normative resources of the law.

To be sure, forays into biological self-fashioning should not be seen as necessarily beneficial for society or the individual. As the Buzzanca case neatly illustrates, parents with access to new reproductive technologies can behave as irresponsibly toward their offspring as bad parents who conceived by natural means. Genetic tests can make people more aware of their health futures and permit wiser planning, but offered carelessly, without counseling or to inappropriate subjects, they can unduly compromise people's healthy enjoyment of the present. And large institutions, such as corporate employers, law enforcement agencies, and private testing companies, retain the power to misuse genetic information for improper forms of social control.

The experiments with life described in this chapter invite us, however, to broaden our conceptions of law and lawfulness beyond the boundaries of formal rules made by courts and legislatures. Indeed, a sense of lawfulness often begins deep inside the laboratory, as scientists struggle to define what is feasible in technical as well as in ethical and even political terms. New forms of expertise, such as bioethics, and new institutions such as ESCRO committees, have arisen to make sure that such deliberation does not occur in overly insulated settings, where science is accountable only to its practitioners' enthusiasm for reliable research, tempered by their internal sense of ethical propriety. Nevertheless, much of society's understanding of what is "natural" does begin with decisions made in labs and clinics about which kinds of biological novelty may be generated without fear, and which others are too risky given current conditions in society. In the next chapter, we turn to a comparable set of questions and puzzles that have arisen from the digital revolution spawned by advances in information and communication technologies.

Chapter 6

INFORMATION'S
WILD FRONTIERS

I f advances in genetics and biotechnology made our bodies
more readable by outsiders, then the new information technologies arguably opened up our minds to parallel forms of
inspection and control. Armed with laptops, iPhones, iPads, and
all the other electronic paraphernalia so abundantly displayed
at airport security counters, we declare our intimate partnership with the thinking gadgets of the digital age. Electronic
machines record our pasts and chart our futures better than fallible human brains ever could, through calendars, photo albums,
music selections, and countless stored documents. Personal electronic devices are partners that we have more or less freely
chosen to live with. Less voluntarily, we daily make our habits
and preferences known to data recorders, from Google to the
National Security Agency (NSA), which can exercise forms of
surveillance and social control that most people only hazily
imagine as they conduct more and more of their business on
the Internet.

By plugging ourselves into cyberspace, we not only gain
access to limitless information but we become information,
making ourselves accessible and potentially vulnerable to new
forms of watching, monitoring, and following. Our patterns

of consumption, social affiliations, images, and even fleeting thoughts expressed on Twitter are promiscuously scattered throughout the digital medium, there to be collected, stored, and pieced together into remarkably comprehensive behavioral profiles of selves that we once considered private and inviolable. And unlike forgiving and forgetful human memories, the memory of cyberspace does not easily fade.

What protections have evolved to guard people against mistreatment or abuse in a world where so much information is so readily available to anyone with an incentive for mining and sleuthing? Whose responsibility is it to ensure that the enormous power and potential of the information age will be used to advance the good of society as a whole and not chiefly to sequester yet more power inside nonaccountable institutions? Answers will depend in part on the flexibility of existing legal and ethical frameworks, such as constitutional law, to stretch and accommodate new imaginations of freedom. The possibilities and constraints of the virtual world may be so different from what has gone before that new forms of association and deliberation and new regulatory instruments will almost certainly be needed.

In this chapter we approach the ethical dilemmas of the information age through several sets of oppositions, asking what is involved when issues are perceived as lying on one side or the other of these divides: digital versus physical selves, public versus private data collection, and the United States versus other nations. Each represents a moving frontier, governed by inarticulate, inconsistent, and shifting norms. Our aim is to tease out who is responsible for making, enforcing, and revising those rules in line with changing social expectations of liberty and privacy in the era of digital personhood.

CONSTITUTIONAL SAFEGUARDS

Let us begin with tradition, in the hallowed preserves of U.S. Supreme Court jurisprudence. In June 2014, the Court handed down one of its rare unanimous decisions on a question of individual liberty. Police searches of cell phones, Chief Justice John Roberts wrote in *Riley v. California*,[1] were illegal unless accompanied by a warrant. Like many landmark constitutional judgments, this one grew from ordinary beginnings: a routine traffic stop in southern California for a man driving with expired registration tags on a San Diego street. On discovering that David Leon Riley also had a suspended license, police officers began a thorough search of his car. Under the hood, they found handguns that they linked to a gang shooting two weeks earlier. Officers also seized Riley's Samsung smartphone and went through its contents, first at the arrest scene and a second time at the police station. Their searches disclosed incriminating pictures and videos that linked Riley to the notorious Crips gang and provided grounds for several felony charges against him, including attempted murder. Riley was convicted and sentenced to fifteen years in prison. There he would have languished, mostly forgotten, except for the fact that a Stanford Law School student clinic picked out his case to build a new argument concerning privacy in the digital era.

Briefs to the Supreme Court must state the legal question the petitioner wishes the justices to address. Riley's was short and to the point: "Whether evidence admitted at petitioner's trial (namely, certain digital photographs and videos) was obtained in a search of petitioner's cell phone that violated petitioner's Fourth Amendment rights." Those rights include protection against "unreasonable" searches and seizures, a provision whose meaning crucially depends on the beliefs of citizens themselves. Did the search violate a zone of privacy of the kind that citizens can reasonably anticipate inside their homes or workplaces or cars and, above all, within their own bodies?[2] To determine the extent and boundaries of that zone, judges must balance the interests of the state against the legitimate expectations of citizens, both filtered of course through their own cultural and professional understandings of reasonableness.

Courts have long since realized that expectations of rights and entitlements change along with developments in science and technology that are forever reconfiguring what we see as normal and, by the same token, what we see as intrusive or alien. Law, especially constitutional law, then becomes an instrument through which a "new normal" gets codified as the thing to defend in the domain of rights. Thus, when blood alcohol tests became widely available, the Supreme Court ruled 5–4 that an inebriated but nonconsenting driver could be subjected to a warrantless blood draw without violating the Fourth Amendment.[3] When law enforcement agencies embraced DNA fingerprinting as the most accurate identification tool available to forensic science, a divided Supreme Court ruled that taking a DNA cheek swab of a suspect arrested on probable cause did not violate the Fourth Amendment, although the search was done without a warrant.[4] Writing for the five-member majority in that case, Justice Anthony Kennedy likened the cheek swab to other routine booking procedures such as fingerprinting and photographing. He focused on the lightness of the swab and not on the depth of information potentially revealed by one's genetic code. DNA tests of arrestees thus became mundane, as an expected and lawful aspect of police procedure, because five justices looking at one test case found it so.[5]

Faced with the warrantless search of a cell phone, however, the Supreme Court justices' imagination and intuitions proved to be far less police-friendly. Chief Justice Roberts noted that cell phones are different from physical objects that officers "reasonably" seize, either to protect their own safety or because they are intrinsically suspect, like powders that could be illegal drugs or cigarettes that do not look like normal cigarettes. Cell phones, he observed, were "now such a pervasive and insistent part of daily life that the proverbial visitor from Mars might conclude they were an important feature of human anatomy." The information they contain poses no risk to the arresting officer; nor can a cell phone help the arrestee escape. Instead, these tiny, portable computers hold massive quantities of different kinds of records—photos, videos, voice messages, Internet search records, electronic mail—that allow a person's entire private life to be reconstructed. Given the sheer scope of digital information packed into a tiny cell phone, Justice Roberts summarily dismissed the government's argument that these are "materially indistinguishable" from any other physical object impounded during an arrest: "That is like saying a ride on horseback is materially indistinguishable from a flight to the moon."

Cell phones, Chief Justice Roberts observed, are hardly even phones: "They could just as easily be called cameras, video players, rolodexes, calendars, tape recorders, libraries, diaries, albums, televisions, maps, or newspapers." It is a remarkable feat of engineering that so much equipment, enough at one time to fill entire rooms, and in the case of books more than rooms, can now be carried inside a purse or pocket, hardly disrupting the tailored silhouette of a slimline jacket. But cell phones do not simply put a world of information at the user's fingertips. As the *Riley* opinion eloquently set forth, they also make their human owners readable to anyone with the time and resources to comb through their contents, such as police officers conducting a search related to an arrest. A cell phone may not be identical with the person who owns it, but it certainly is a repository of much that makes the individual owner recognizable, identifiable, and unique.

A person who shares information and thoughts with a thinking machine does not have the same perceptions of selfhood as a person who has never encountered such a technology. A number of years ago, I relocated from one university to another. Moves are always difficult, and watching a well-loved house get stripped of its contents is wrenching even if one is looking forward to nesting somewhere else. But for me the sharpest pang of separation came when I unplugged my office computer for the last time late the night before we left. That was when it sank in that I would be leaving behind not just a place, a home, and a network of friends, but an aspect of the scholar and person I had been. It felt like closing down a part of my consciousness. It is important to ask to what extent our institutions of ethical, legal, and social analysis are capable of taking on board these sorts of subjective changes that come about through our interactions with technologies that have become, in some sense, our most intimate companions, extensions of our minds and of our potential selves, as we fashion those selves now and in the future.

PHYSICAL AND DIGITAL SELVES

In the imagination of the computer scientists and engineers who invented it, cyberspace was at first more a territory than an aid to self-fashioning. It was supposed to be, above all, a place without constraints, where people would be able to express themselves with a freedom they lacked in grounded, earthly institutions. Access, to start with, cost virtually nothing—at least to those who understood how to claim a piece of the digital pie for themselves. Buying a page in a major daily newspaper such as the *New York Times* was, for most people, prohibitively expensive. The Internet reduced those barriers to near zero, and as new tools developed even programming skills became increasingly irrelevant to placing one's content on the Internet. Space, too, was plastic and infinitely available, not hemmed in by the physical limits of paper or bandwidths or cement. Today, users with only modest amounts of computer savvy can set up a website on a server that costs very little and begin broadcasting their views or displaying their wares to any who care to browse their offerings.

That freedom, however, comes with a potential for control that the pioneers of virtual space did not entirely foresee. Digital transactions, especially in the interactive Web 2.0 era, are two-way streets. By putting oneself into cyberspace, one opens the way to pervasive and perpetual observation. As the intelligence contractor Edward Snowden revealed in 2013 through his massive leak of NSA documents, our electronic communications are not shielded against warrantless searches by the government to the extent U.S. citizens once believed them to be. Nor is the state the only watchful actor. A giant commercial marketplace like Amazon places the goods of the world at the buyer's fingertips. In return, however, Amazon gets to record an entire history of purchases and preferences, to aggregate it over time, to sort it with algorithms, and to create a picture of each user as a potential target for advertisers and sellers of commodities beyond those the user originally accessed.

Google, similarly, maintains records of each user's searches; and, as we will see below, it can and does scan the e-mails of users of Gmail for illegal content. Users, moreover, no longer simply seek information and become unwitting data sources; they also share information intentionally via social networking sites like Facebook, Flickr, YouTube, Pinterest, and Twitter. The range of people one shares with in the free-flowing medium of cyberspace is not wholly within the sharer's control. To adapt a phrase used by the psychologist and Internet critic Sherry Turkle, the electronic devices whose ubiquity Chief Justice Roberts noted in *Riley* are "always on/always on you."[6]

Whether one regards the increased transparency of the digital self as problematic, even abusive, depends on one's social and political preferences, with age, gender, and place of origin playing important defining roles. The Harvard law professor Cass Sunstein, for example, claims that consumers want to have their minds read and their wishes catered to through "predictive shopping" even before they manage to articulate their needs for themselves.[7] Not all such readings of human minds can be considered benign, however; nor do all consumers react the same way when choosing between liberty and efficiency in the marketplace. Protections are needed for digital selves, but where should the safeguards come from, and to what extent can the wine of old principles be poured undegraded into the new bottles of the digital age?

Laws protecting individuals against unwanted intrusions developed in a physical world, and the analogical reasoning that courts like to use still draws primarily from the assumptions of that world. A physical imagination underwrites the Fourth Amendment's guarantee of the "right of the people to be secure in their persons, houses, papers, and effects." Both common people and judges imagine unlawful intrusions in terms of government officials kicking open the doors to their homes, seizing their possessions, or assaulting their persons without just cause. Such fears, for instance, led the U.S. Supreme Court in 1965 to strike down a Connecticut state law forbidding the prescription of contraceptives to married women.[8] Writing for a 7–2 majority, Justice

William O. Douglas concluded that the Constitution protects a right to "marital privacy," safeguarded within the home. Hence, he likened the ban on contraceptives to a physical intrusion into the marital bedroom and the sanctity of the age-old relationship within. Couples engaged in this "harmony in living" and this "bilateral loyalty," Justice Douglas decided, could not be denied access to counseling in the use of contraceptives.

Similar analogies allowed courts to extend protection to people's nonphysical traces even before a large fraction of the world's population acquired digital profiles. Thus, people have a right to control the dissemination of their name, their image or likeness, and even their signature gestures or mannerisms under a general "right of publicity" or "personality right." The status of these markers as property will be discussed in chapter 7. Here it is important to note that such virtual extensions are regarded as components of personality and selfhood as if they were part of the physical person. The gesture or the signature (John Hancock's on the Declaration of Independence, for example), let alone the physical likeness, is so closely associated with an individual that it serves as a metonym, instantly conjuring up the whole person and hence deserving protection in the same way.

Legal protections also extend to physical places that people ordinarily regard as closed to outsiders, such as bedrooms, locker rooms, or fitting rooms, and many states treat the making of unauthorized voyeuristic images or visual recordings in such spaces as a criminal offense. Such laws extend the rights a person enjoys as a spatially located, physical being—liberty, autonomy, privacy, bodily integrity—to representations that are deemed integral to one's inviolable self. But is the analogical mapping between physical and virtual adequate to meet the challenges confronted by our surrogate digital selves? That question requires closer attention to the specific properties of the selves we are increasingly committing to the Internet. Answers depend on how we conceptualize the relationship between our real and our virtual identities, and that is a contested and evolving process.

NEW VULNERABILITIES

Digital traces of persons differ from physical ones in ways that make it more problematic and less useful to draw straightforward analogies with physical persons or spaces. First, there are novel issues of individual autonomy. "Big data" have become big business in the era of Web 2.0 in part because large data sets allow access to hitherto secluded features of human thought and intention. Digital data from countless online interactions can be aggregated and grouped to build remarkably accurate, composite pictures of a person's identity, attitudes, and behaviors; data allow access not just to that person's body or appearance but to the thinking and acting self within. Second, digital information is reciprocal, raising new questions about privacy: persons not only put data about themselves onto the World Wide Web but, through those acts, also become potential subjects for surveillance, commerce, and even experimentation. And as people and their actions are hooked up to myriad data collection devices in the Internet of things, they also become traceable as never before. Third, unlike physical movements, informational traces are lasting, some might say unforgiving and even accusatory. They persist through time, sedimenting histories that are less amenable to personal control or erasure than lives lived in the predigital era. All three dimensions allow outsiders access to aspects of the personality that could not be deduced from physical traces alone; all three present dilemmas for ethics and law.

The Internet has become an informational quarry. Examples

of people being tracked, or tracked down, through publicly available information are legion. A striking example, a first of its kind, dramatized the inroads such data allow into presumed zones of privacy. Sometime in late 2004, a fifteen-year-old boy conceived with sperm from an anonymous donor set out to find his genetic father. He "rubbed a swab along the inside of his cheek, popped it into a vial and sent it off to an online genealogy DNA-testing service." Nine months later, for a fee of $289, the service identified two persons with closely matching DNA who revealed themselves to the boy as likely near-relatives. Using their shared last name and the father's place and date of birth, which he knew from his mother, the boy then bought from an Internet genealogical service a list of persons meeting all of his criteria. One person on the list had the right last name, and so the boy finally had his answer. The man had never donated his DNA to a database or consented to having his identity as sperm donor made public; nonetheless, a fifteen-year-old's ingenuity and persistence outed him, at very little cost to the inquirer. "The case shows," said a Canadian bioethicist, "that there are ethical and social concerns about assisted reproduction that we did not think about."[9] The comment characteristically recasts the privacy breach as an unintended consequence, but of course the easy availability of data is a conscious design feature, not a bug, one of the attributes fought for and prized by the originators of cyberspace.

The fifteen-year-old used ingenious common sense to find his father. More sophisticated computational techniques can be applied to publicly available data from the digital environment to reveal personal traits that people had no intention of making public. For example, a University of Cambridge study of 58,000 volunteers, published in the prestigious *Proceedings of the National Academy of Sciences* (*PNAS*), showed that a mathematical analysis of their Facebook Likes could correctly discriminate "between homosexual

and heterosexual men in 88% of cases, African Americans and
Caucasian Americans in 95% of cases, and between Democrat
and Republican in 85% of cases," as well as between Christians
and Muslims in 82 percent of cases.[10] The authors worried that the
threat of exposure might deter people from using digital technologies, but they still expressed a Pollyannaish hope that increased
transparency and control over information would ensure adequate
trust and goodwill between technology providers and users.

As things currently stand, it is difficult to engage in any form
of digital behavior without opening oneself up to some degree as a
subject of investigation. People who use search engines and online
stores or services must generally agree to some sacrifice of privacy,
through reciprocal arrangements whereby the provider can use
their e-mail addresses and other information. Some, but not all,
providers promise not to share users' private information, but how
far they go in keeping their promises is hard for users to monitor, and a lively global hacker culture intent on identity theft poses
additional threats. There have been many well-publicized attacks
on business sites, such as the Sony hack of Christmas Eve 2014
that revealed masses of confidential information about the company's internal management practices. Few attacks were more blatant
or embarrassing than the hacking of the extramarital affairs site
Ashley Madison in July 2015, which revealed e-mail addresses and
account information about some thirty-seven million users.[11] The
hackers claimed their actions were partly aimed at deterring a
company whose practices they disliked, but their actions inflicted
damage on people, such as spouses and children of the site's users,
who were in effect innocent bystanders. There were even reports
of a few suicides linked to the disclosure of names.

Even in the absence of a deliberate cyber attack, social media
use erodes personal privacy, often without users' knowledge or
consent. Facebook, the company that launched the social media

revolution, has drawn so much protest and even legal action over its relatively short life that Wikipedia devotes an entire page to "Criticism of Facebook." Its privacy policies have repeatedly stirred up controversy, for example, when the company decided that it would not allow accounts to be permanently deleted from its servers[12]—a policy Facebook subsequently reversed.

More insidiously, the two-way street of Web 2.0 makes possible a kind of psychological experimentation that goes beyond older propaganda campaigns indiscriminately aimed at the masses. Many companies conduct ongoing research on customer data, ostensibly in an effort to improve the products they offer and to tailor them more carefully to individual users' needs. These practices are generally accepted as legitimate adjuncts of running a business. Internet service providers, too, conduct research with customer data for better messaging. Consent to having one's information used for such "operational" purposes is often a condition for using a site, although it may be given in fine print that users have not troubled to read. While such uses may be unpalatable to many, especially when it involves sharing information with other service providers, arguably it is an allowable extension of the original contractual relationship. The line between permissible and unethical research, however, is blurry and companies—in a virtual vacuum of regulatory controls—can and do overstep.

Facebook users could not have imagined that their activities on the site would turn them into psychological study subjects. Yet in early 2012, Facebook conducted a mood experiment on almost 700,000 anonymous users without their knowledge or consent.[13] Published in the reputable scientific journal *PNAS* in June 2014,[14] the study reported that Facebook had purposefully altered the emotional content of newsfeeds sent to users for one week, tuning the content to be either more positive or more negative through the expedient of "deprioritizing" messages with emotionally negative words for a part of the study population. People receiving relatively positive stimuli produced on the whole somewhat more positive posts, while the control group of people receiving more negative stimuli did the reverse. The three authors, a Facebook data analyst named Adam Kramer and two Cornell University statisticians, concluded that this was evidence that emotional contagion can spread through a network even without person-to-person contact.

A national outcry followed. The study had not crossed any legal lines, but media commentary on the researchers' ethics was overwhelmingly negative. Only Facebook's internal reviewers had cleared the study in advance. Since the Cornell researchers had not collected the human subjects data themselves, but were evaluating an external data set, the university's institutional review board (IRB) did not see any need to approve their involvement in analyzing the data—even though the paper's *PNAS* editor, the Princeton psychologist Susan Fiske, had assumed it was somehow IRB approved. Kramer himself felt the need to post an explanation and apology on Facebook: "I can understand why some people have concerns about it, and my coauthors and I are very sorry for the way the paper described the research and any anxiety it caused. In hindsight, the research benefits of the paper may not have justified all of this anxiety."[15] Some ethicists felt that the matter was being overblown, but not everyone found Kramer's after-the-fact response satisfactory. Thomas Ohm, a University of Colorado law professor and longstanding critic of power imbalances on the Internet, observed, "The ethics have been begging to be discussed. There's A/B testing to better deliver a product a customer wants, but it's another thing for companies to consider users to be a willing and ready pool of lab rats that they can prod however they want."[16] Despite spirited exchanges on the Internet, the task of resolving whether a corporate-sponsored study treats people permissibly as customers or impermissibly as lab rats rests at present largely with the designers. The ethical apparatus for scrutinizing such studies in advance simply does not exist.

Time and memory, too, work differently for physical and digital selves. Bodies die and memory disintegrates, but digital data can live on indefinitely. It is difficult enough for ordinary people to clean their own personal computer's hard drive of all its obsolete content. It is far harder to eliminate information scattered through the vast reaches of cyberspace. Pictures posted to amuse friends on Facebook have cost people jobs years later when prospective employers saw them and found them compromising or unseemly.[17] Twitter, a medium aiming to capture one's most informal and evanescent thoughts, turned out to be a site that is both glaringly public and, for all practical purposes, permanent as well. One may be dogged for years or decades by tweets sent during a relationship meltdown or a crisis at work. Companies like Facebook have gone back and forth on policies regarding whether personal data, once posted, can be deleted. New commercial services have sprung up purely for the sake of tracking down and deleting information one no longer wishes to leave in the public domain. As yet, however, rules governing the degree to which one can regulate the life span of one's own data remain fluid and, as we see below, also cross-nationally divergent.

PUBLIC OR PRIVATE

The pioneers of the information age celebrated freedom from what they saw as the stranglehold of nation-states exercising their sovereignty across physical territory. Borrowing a term coined by William Gibson, author of *Neuromancer*, John Perry Barlow, onetime lyricist for the Grateful Dead and an early

Internet guru, drafted a defiant Declaration of the Independence of Cyberspace: "Governments of the Industrial World, you weary giants of flesh and steel, I come from Cyberspace, the new home of Mind. On behalf of the future, I ask you of the past to leave us alone. You are not welcome among us. You have no sovereignty where we gather." Prophetic and utopian, his words resonated with a generation of youthful computer wizards flushed with the triumph of discovering a new and seemingly boundless land, where all could enter as equals and all forms of speech and expression would be free.

Today we know better. In the wake of Bradley (later Chelsea) Manning's thirty-five-year prison sentence for leaking classified U.S. documents to WikiLeaks, the prolonged sheltering of the WikiLeaks founder Julian Assange in the Ecuadorian embassy in London, and Edward Snowden's seeking and finding refuge in Russia to escape prosecution as a traitor to the United States, few would question that state sovereignty is alive and well in regulating content on the Internet—at least when that content pertains to national security as defined by nation-states. Far from being the space of total personal freedom contemplated by Barlow and other pioneers, it has emerged as a space crisscrossed by lines of power that enmesh those who enter it in invisible and unpredictable ways. Both public and private entities control individual liberty in cyberspace but through different mechanisms and under different rules of transparency and accountability.

No Global Norms

Given the enormous publicity generated by Snowden's revelations, it is remarkable that world opinion remains so sharply divided on whether he should be seen as hero or as traitor, whistleblower or criminal. Part of the reason does indeed have to do with the dissolution of territoriality as it was known in the 1970s, and the corresponding weakening of the powers of state institutions. The power to draw public opinion together under such slogans as loyalty or national security has eroded, but that fragmentation is not only, or even mainly, the result of cyberspace emerging as a "new home of Mind." Other technologies of travel and dispersal also matter, and the electronic media play an important role.

The differences between the cases of Snowden and Daniel Ellsberg, who previously held the record for disclosing American state secrets, are both fascinating and instructive. Ellsberg was an antiwar activist and Harvard-trained decision theorist who surreptitiously copied the Pentagon Papers, a top-secret report prepared for his boss, Secretary of State Robert McNamara. Ellsberg provided the documents to the *New York Times* and other newspapers. The *Times*, later joined by the *Washington Post*, began publishing articles based on the papers in June 1971 and was promptly taken to court by President Richard Nixon's Justice Department. The administration sought a restraining order against the *Times* and the *Post*, arguing that continued publication of the Pentagon Papers would damage U.S. foreign relations and aid the nation's enemies. The Supreme Court ruled for the press in a case widely seen as a resounding victory for the First Amendment's free speech principle.[18] Ellsberg was brought to trial under the Espionage Act, but the presiding judge dismissed the charges after revelations of grave misconduct, such as illegal wiretapping, in the government's acquisition and handling of evidence. Ellsberg became something of a folk hero, continuing to speak out and demonstrate against violations of free speech well into his eighties.

The story of the Pentagon Papers played out within a single nation's territorial jurisdiction, involving primarily its citizens, its news media, its state secrets, its laws, its law enforcers, and its judges. The Snowden case, by contrast, was legally, politically, and physically dispersed from its origins, even though the centerpiece—as in the Ellsberg case—was a U.S. citizen's theft and disclosure of classified U.S. government records. Snowden himself was living and working as an NSA contractor in Hawaii when he flew to Hong Kong to meet with two Americans, the documentary filmmaker Laura Poitras and the lawyer-journalist Glenn Greenwald, the former residing in Berlin, Germany, and the latter in Rio de Janeiro, Brazil. Greenwald lived in Brazil purportedly because the American Defense of Marriage Act would not allow his same-sex partner to live with him in the United States. Poitras and Greenwald worked for a British paper, the *Guardian*, which broke the NSA surveillance story on June 5, 2013; four days later, Snowden came forward to reveal himself as the face behind the leak. On June 23, Snowden flew from China to Russia, where he remained as a refugee of uncertain status and a fugitive from justice in the eyes of the U.S. government, which had revoked his passport. His disclosures unleashed a widening tangle of events—among them the *Guardian*'s demolition in July 2013, at the UK government's insistence, of the computer hard drives containing the leaked documents; Snowden's televised testimony to the European Parliament in March 2014; and the award of a 2014 Pulitzer Prize to the *Guardian* and the *Washington Post* for public service reporting by Greenwald, Poitras, and Ewan MacAskill.

All of these episodes reveal a world densely interconnected by the technologies of modernity—planes, phones, television, computers, e-mail, electronic newspapers—but with those connections still overlaid on a bedrock of national rivalries and competing sovereignties. Relations between the United States and Latin America soured when a plane carrying the Bolivian president Evo Morales was forced into an unscheduled landing in Austria in July 2013 on the suspicion that he might be transporting Snowden away from Russia. Reports that the NSA had tapped Chancellor Angela Merkel's private telephone, based on documents disclosed by Snowden, infuriated Germany and Europe and strained relations between Merkel and President Barack Obama at a time of heightened global crisis.[19] Meanwhile, Snowden continued to be seen by many as a standard-bearer for liberty. In early 2014, he joined the board of directors of the Freedom of the Press Foundation at the invitation of its cofounder Ellsberg and others, including Greenwald and Poitras.

Snowden's audacious actions spurred a global public debate on surveillance, with high-level political repercussions around the world. It embarrassed world leaders and security agencies, spurred legislative inquiries, prompted constitutional reflections in the United States, intimidated but also rewarded the press, and made publics in democratic nations aware of the vast and unsuspected potential for abuse in the security apparatus of their own states. At the same time, in marked contrast with the Pentagon Papers case, the Snowden affair and its multiplying knock-on effects also illustrated the difficulty of reaching closure on the issues that Snowden had forced into the open. In the post-9/11 world, many felt that U.S. constitutional law no longer protected freedom of speech or those who, like Ellsberg, might once have risked prosecution in the hope of vindication before an impartial tribunal. Still less were institutions in place to adjudicate the ethics of surveillance in the age of metadata sweeps and intercontinental telephone tapping. Put differently, the Ellsberg saga played out in a well-developed national public sphere, where the rules of debate and the language of public reason were relatively well grounded in established institutional practices.[20] The Snowden case embroiled the world community in a multicentered debate that could not have been imagined fifty years back, but it also underlined the difficulty of forging collective norms across a globe sharply divided by national interests and cross-national political divisions and alignments.

Data Oligarchs

Tech billionaires emerged as a social class only in the early years of the twenty-first century, but the concentration of extraordinary wealth in a small number of hands, based mostly in California's fabled Silicon Valley, is emblematic of a new resource—personal information—that came to be controlled by a relative few. The term "data oligarch" has been applied to companies such as Google and Facebook. They open unparalleled informational gateways to the masses but they also control masses of information whose very scope and variety give them enormous potential value. Though technically private, or nonstate, these companies straddle a line that makes it difficult for states to hold them accountable for data use, or misuse.

Google offers a particularly interesting case study in the interplay of power and accountability in cyberspace. Google started out trying to be the best search engine ever created, dedicated to giving its users correct and unbiased information on whatever they were searching for, and doing it speedily. The company's famous motto, "Don't be evil," sounded a theme of social responsibility and overall commitment to good behavior that is one of Google's trademarks. In its ventures in China, for example, Google stood up against censorship at the risk of periodic confrontations with and shutdowns by the Chinese government. As the company grew, however, its products diversified and its ambitions expanded. Acknowledging these developments, Google in 2015 formed a new parent company called Alphabet to encompass all of these diversified activities; interestingly, Alphabet did not pick up the "Don't be evil" slogan as part of its code of conduct.[21]

No longer simply the world's most popular search engine, Google also markets a host of Internet products and services such as Gmail, Google Maps, YouTube, the Chrome browser, as well as the Android telephone. Its profits exceed $50 billion per year. Google now behaves in some respects like a state, bringing information and advertising to a population of more than a billion searchers each month, thus acting on a scale comparable to that of a large national government. But does Google use its power in the beneficent fashion called for in its original code of conduct?

Google's frictions with both the U.S. government and the European Union point to problems. The company has for years been investigated for antitrust violations stemming from charges that it uses its huge market share to disadvantage competitors unfairly— for example, by displaying search results that favor its own mapping and finding services. In a U.S. Senate antitrust hearing in 2011, Google's CEO, Eric Schmidt, admitted that his company's market share looked monopolistic, but that this was ultimately an issue for the courts, "I would agree, sir, that we're in that area. . . . I'm not a lawyer, but my understanding of monopoly findings is this is a judicial process."[22] Although U.S. investigations did not lead to serious restrictions or fines, an eventual European settlement could go much further.[23] In the wake of Snowden's revelations, EU regulators became more then ever concerned about American companies dominating information collection and flow within Europe's territorial borders.

Controlling the kinds of information that people see when they search is a new and unregulated kind of private power. Formerly, it was primarily the nation-state that regulated information flow, through its control of public education and, to greater or lesser extent, the news media. Indeed, the political theorist Benedict Anderson argued that nationalism itself was the product of such top-down control, creating "imagined communities" of people who were induced to see themselves as belonging to a single nation.[24] Yet to say that Google sets people free from state-owned and state-controlled media, as for example in China, clearly is not the whole story. As a private concern with profits in mind, the company, no less than a nation-state, has a huge interest in controlling what its users see, and even how they think and act, by selectively revealing and withholding information. As Cass Sunstein's comments indicate, such mind control may not be unpalatable to some consumers so long as it is dressed up in the language of choice, like "predictive shopping." But Google exerts more direct and less benign control over people's lives, beginning with its employees.

Soon after the death of Steve Jobs, the iconic head of Apple, more than sixty thousand technical workers filed a class action lawsuit charging Apple, Google, and other prominent high-tech companies with antitrust violations in their hiring practices.[25] The companies, plaintiffs alleged, had entered into express and tacit agreements not to poach each other's skilled employees or to enter into wage competition with one another. Jobs was a key architect of this state of affairs; his mythic status in Silicon Valley helped persuade others to fall in line, and of course lower wages and less competition benefited all employers collectively. The pattern of collusion allegedly cost the workers $3 billion in lost compensation. Three years after the lawsuit began, Judge Lucy Koh of the U.S. District Court in San Jose, California, rejected a proposed settlement of $324 million as wholly inadequate. She stressed that a jury would find the evidence presented against Jobs and his colleagues compelling, leading to potential liability many times the proposed amount. Observers thought that a sum in the range of $1 billion, threefold the original proposal, would be closer to what the judge might find reasonable under the extraordinary circumstances of this case.[26] In the end, the judge approved the considerably more modest amount of $415 million, still one-third more than the companies' initial proposal.

The tendency to make and live by their own rules that marked the tech companies' relations with highly skilled employees spilled over into troubled community relations in the Bay Area. Google makes a point of tailoring its workplaces to cater to employees' mental and physical well-being, paying attention to everything from fitness centers to the colors of its walls. Those private perks, and the huge influx of wealth they mirror, created rising disparities between the upwardly mobile tech sector and the grittier urban environment that gave them their start. Unexpectedly angry demonstrations broke out in San Francisco in early 2014 against the buses that ferry employees to the fabled "campuses" of Google, Apple, Yahoo, and other companies—buses that some said were illegally using city bus stops.[27] In a placating gesture, Google launched its first Bay Area Impact Challenge, inviting local nonprofits to compete with bright ideas for community improvement. Winners selected by the community during a two-week public voting period received a total of $5 million, with the four top applicants winning awards of $500,000 each. Notably, this gesture mimicked classically democratic processes of spending on public works and seeking electoral validation (191,504 "votes" were cast from May 22 through June 2), but without the conventional trappings of democratic deliberation, such as hearings or active efforts to involve citizens.

Google, the data gatherer, functions even less like a private entity when it enters into partnership with governmental authorities, as it has done in certain areas of law enforcement. The company routinely scans Gmail users' e-mails for evidence of child pornography, a crime under federal law. The National Center for Missing and Exploited Children maintains a database of digitized and encrypted images of missing or victimized children. If one of these shows up when sent on Gmail, Google reports it to authorities. Acting on such a tip, Texas police in August 2014 searched electronic devices belonging to a restaurant worker who had been convicted of sexual assault twenty years earlier. The discovery of pornographic materials and related messages led to the man's arrest. While few considered the identification of a potential child abuser problematic, observers noted that the same sort of technology that Google uses in detecting child pornography could be used to obtain other kinds of information that might lead Google down a slippery slope, acting in effect as surrogate eyes for the state on matters that may not properly be the state's business.

The extent of Google's and other high-tech companies' involvement with PRISM, the government's secret electronic surveillance program, also caused public consternation. Snowden's leaks revealed that the NSA, in collaboration with Britain's chief intelligence agency, was collecting vast amounts of daily information on the electronic communication patterns of Americans, as well as foreigners. Those "metadata" include records of most telephone calls, including the numbers of the caller and recipient and the call's duration. Also included in the NSA's data sweep were e-mails, photographs, audio and video chats, and documents. Legally, the PRISM program operates under the authority of the Foreign Intelligence Surveillance Act (FISA) of 1978. Except under specified circumstances (e.g., when monitoring lasts less than a year), the NSA needs an authorizing court order from the Foreign Intelligence Surveillance Court (FISC). The FISC, its seven judges appointed solely by the Chief Justice of the United States, almost never denies the NSA's requests for warrants, although the court's actual decisions are shrouded in secrecy.

In one of their first articles for the *Guardian* based on Snowden's leaked documents, Greenwald and MacAskill reported that the NSA had obtained direct access to the central servers of Google,

Facebook, Apple, Yahoo, and several other U.S. Internet companies.[28] Google, however, categorically denied that the NSA had either direct or indirect access to its data systems and, together with the other tech giants, insisted that it discloses data only case by case and after thorough internal review. In September 2014, documents unsealed by the appellate court that reviews FISC warrant denials revealed that in 2008 the U.S. government had threatened Yahoo with fines of $250,000 per day if the company refused to hand over communications data under the PRISM program.[29] Yahoo challenged the constitutionality of the law demanding bulk electronic data, but that effort failed before the FISC.

Briefly, in the winter of 2016, it appeared that another test case would pit a giant of the tech industry against the United States. The Federal Bureau of Investigation (FBI) asked Apple to unlock the iPhone of one of the shooters in the terrorist attack that killed fourteen coworkers at a pre-Christmas office party in San Bernardino, California. Apple resisted, arguing in part that compliance would violate its First Amendment rights by forcing it to write new software, but also that meeting the government's request would break down the company's system of password protection, rendering all iPhone users potentially vulnerable.[30] When talks with the FBI stalled, Apple's CEO, Tim Cook, took the company's case to the American public, declaring in a February 16 statement, "The FBI may use different words to describe this tool, but make no mistake: Building a version of iOS that bypasses security in this way would undeniably create a backdoor. And while the government may argue that its use would be limited to this case, there is no way to guarantee such control."[31] The prospect of an ugly legal battle melted away when the Justice Department announced some weeks later that it had found a secret way, without Apple's help, to hack the phone in question. The case underscored not only some huge, unresolved, technical questions about data security but, more importantly, about whether citizens should count on the state or on private industry for protection in an era when digital devices resemble, in Chief Justice Roberts's words, "an important feature of human anatomy."

THE UNITED STATES IN THE WORLD

In the fallout from Snowden's disclosures, few allegations proved more divisive than the report that U.S. intelligence had tapped into Chancellor Merkel's private telephone. It drove a wedge of distrust into relations between two traditionally close allies, Germany and the United States, at a time in world history when Western unity of purpose seemed crucial. Almost a year after the report surfaced, Merkel indicated to Obama during a state visit to the United States that the two nations still differed in their ideas of "proportionality" when balancing security against privacy.[32] It was as if by tapping the phone of Europe's most powerful leader, the United States had also tapped into a deep vein of transnational discord about the ethical and allowable conduct of international relations in the electronic age.

Europeans, chastened by their experiences of fascism and socialism during World War II, are often said to care more deeply about privacy than U.S. citizens. Inside the U.S. Holocaust Memorial Museum in Washington, D.C., almost the first exhibit one sees is a Hollerith machine, a precursor of the modern computer used in carrying out the 1939 German census that for the first time recorded citizens' race along with other personal information. The rejoinder is that Americans care more about security, especially with their minds attuned to terrorism after 9/11. While such flat generalizations should be handled with care, substantive policy divergences point to significant differences in how the state imagines its relations with digital citizens on the two sides of the Atlantic.

The European Union (EU) has taken steps since the 1990s to create a uniform framework for personal data protection across all of its member states. The Data Protection Directive of 1995 established a set of rights that the "data subject" can assert against a public or private "data controller." A data subject is a natural person (i.e., human being) who is identifiable by means of a number or by specific features of physical, mental, economic, cultural, or social identity. The directive provides that the data subject has a right not to have his or her data collected or "processed" without explicit informed consent, except under specific conditions, such as for the subject's own protection or for public functions such as tax collection. Since 2012, the EU has been working to replace the current directive, with its patchwork of national implementation systems, with a single regulation that will apply as law throughout the Union. While streamlining the data protection process for companies, the new law also aims to strengthen privacy provisions for individuals, such as the principle of express prior consent to data collection or processing, and an explicit "right to be forgotten."

The Court of Justice of the European Union preemptively addressed the latter issue in May 2014.[33] Mario Costeja González, a Spanish lawyer, complained to Spain's data protection agency in 2010 that Google searches for his name were linking to two pages from a large Spanish daily newspaper reporting in 1998 on forced sales of his property because of an earlier failure to pay social security debts. Costeja insisted the pages were "no longer relevant" and demanded that they be altered or suppressed and that Google and its Spanish subsidiary remove the links. On reviewing Costeja's claim, the Court of Justice ruled that even data once lawfully collected and stored might over time become incompatible with data protection mandates where "the data appear to be inadequate, irrel-

evant or no longer relevant." The court noted, too, that the right
to have data deleted is not absolute but that a "fair balance" should
be struck between the data subject's rights and the public interest.

The ramifications of the ruling have yet to sink in, but it was
clear within months that the process could become immensely
burdensome for Google and other big data controllers. Google announced that it would review the tens of thousands of
requests that soon came pouring in for compliance with the
court ruling. That, however, would require a close look at the
nature of the challenged data to see whether they were indeed
outdated or irrelevant, as well as a determination of the public's
interest in having access to the information. Old information
about a public figure or someone seeking a position of authority,
for instance, might be deemed "relevant," whereas indefinitely
maintaining purely private information, such as the record of
Costeja's decade-old bad debts, might be deemed "excessive."

The right to control the long shadow of one's digital past is an
important new principle of data protection, but time will tell how
efficacious that principle proves to be. What are the economic and
social implications, for instance, of divergences between the EU
and the United States on a principle such as the right to be forgotten? This is not a hypothetical question. Early reactions displayed
considerable differences of opinion. Jonathan Zittrain, a Harvard
law professor, told the *New York Times*, "I think it's a bad solution
to a very real problem," whereas Viktor Mayer-Schönberger, professor of Internet governance at Oxford, said the ruling merely
affirmed existing law and welcomed a return to "the ephemerality
and the forgetfulness of predigital days."[34]

For the present, as with agricultural biotechnology (see chapter 4), the two powerful political and economic zones on either
side of the Atlantic appear to be following divergent conceptual
and regulatory pathways, consistent with each one's longstand-

ing convictions about good relations between states, markets, and individuals. The United States has no overarching law on data privacy such as the EU directive or the proposed regulation. Instead, a patchy system of protections ranges from a high national standard for medical records under the Health Insurance Portability and Accountability Act to virtually no protection against the collection and processing of data on consumer preferences or voluntarily uploaded personal information. This state of affairs reflects a U.S. penchant for industry self-regulation, with national law stepping in to close loopholes only after they rise to political salience, tempered mainly by a national preoccupation with privacy in matters pertaining to health, health care, and the possible denial of medical insurance.

LIVES ON THE EDGE

The expansion of the electronic frontier has changed the way much of the world lives. Gone are the old-fashioned rituals of writing Christmas cards, mailing letters, reading books, and, for a growing population, making out checks. To American millennials, in particular, any transaction calling for paper feels so twentieth century, a messy relic of a bygone age. Even the sound of the human voice, telephonically conveyed, has become a bit outmoded, as people "talk" on Facebook or text and tweet their connectedness with friends. Distant contacts have come closer, while traditional meeting places, like the post office, the neighborhood bookstore, the county library, and even the local shopping mall, die slow and unloved deaths. California's storied Silicon Valley lures a gold rush of young entrepreneurs eager to mine the seemingly unlimited wealth of digital space.

The unseen by-product of all this ferment is the appearance

of a digital alter ego for every individual with a modicum of Internet presence. Habitual users of search engines and social media are in some respects more knowable through their digital traces than through old-fashioned face-to-face contact. It is the Internet that records, and maybe saves forever, their passing thoughts, their casual photos, their accumulated writings, their speech, their purchases, and, in some fraction of cases, their dark, shady, or criminal impulses. The lords of this unruly universe are not only governments, although, contradicting initial hype, state sovereignty remains a powerful force on the Internet. Paralleling states in their ability to "read" people and control human behavior, however, are the new data oligarchs: Google, Microsoft, Apple, Amazon, Facebook, Twitter, Yahoo, YouTube, and others not yet so well-known or omnipresent.

Internet governance is now a well-recognized policy domain, but those who see it primarily in terms of pricing access to the Internet miss the enormous ethical and legal dilemmas associated with governing the virtual and undying subjects and populations that have been generated by real people acting in real time with digital resources. Legislatures have been slow to respond, hampered in part by the fear of inhibiting economic growth and technological development. Courts, their imaginations constrained by analogies to the physical world, have delivered partial and inconsistent judgments. In this context, the EU ruling on the right to be forgotten, for all its impracticalities and imperfections, stands out as a beacon pointing the right way: toward new and creative reformulations of what it means to be not only human in the twenty-first century but also a moving, changing, traceable, and opinionated data subject.

Chapter 7

WHOSE KNOWLEDGE, WHOSE PROPERTY?

P ersons have rights in law and ethics that no one can violate or remove. We hold human persons to be integral beings, possessing dignity, with rights to guard their physical selves, belongings, and immediate surroundings against illegitimate attack or intrusion. These rights of personhood are treated as inviolate and inalienable in a wide variety of legal and ethical frameworks, from national constitutions to the United Nations' Universal Declaration of Human Rights. In these respects, persons are notably different from things, especially those things that can be owned, or property. Things we own can be used or used up, sold, bartered, exchanged, divided up, given away, or destroyed at the owner's will, unless explicitly prohibited by law. New biological and information technologies, however, have muddied the line between persons and property, by allowing persons in effect to distribute aspects of their selfhood in ways that were once unthinkable. What is the status of materials derived from our bodies and selves, whether these are physical entities such as genes or digital records of our words and transactions? In this chapter we consider some of the quandaries presented by the technological divisions and extensions of personhood, and the emerging responses to those challenges.

IMMORTAL CELLS

It could be a riddle from a modern-day Sphinx. What is dead but still lives, is named and yet nameless, has harmed one life but could save many? The answer, found in physical form in any contemporary biology lab seeking to make drug discoveries, is a human "cell line"—a population of cells extracted from a mortal, often diseased, person's body that can, under appropriate conditions, live forever, supplying endless material for research leading to treatments for humankind's most frightening and debilitating illnesses.

Normal cells have a finite life, after which they die. The cells in a cell line, however, have undergone a mutation that keeps them dividing indefinitely even outside the human body, becoming in effect immortal. These proliferating, genetically identical cells are an indispensable tool in modern biomedical research. Not only do they perpetuate themselves, but they can be multiplied many times over through cloning. Scientists can use these bountiful resources to conduct studies that could not be done with scarce materials. By testing promising therapies on cell lines, researchers also avoid exposing living humans to potentially toxic side effects that could accompany any experimental drug. Of course, medications must eventually be approved for human use through trials involving real people, but the use of cell lines at the outset makes it possible to separate out the more promising avenues from less good alternatives without harming anyone in the process. But who owns the cells and the genetic information derived from a person's body, who decides what can be done with them, and who shares in the profits if a successful treatment emerges?

In August 2013, Francis Collins, director of the National

Institutes of Health (NIH) and Kathy Hudson, deputy director for science, outreach, and policy, addressed some of these very basic questions in relation to an extraordinary case. In a comment in *Nature*, Hudson and Collins described an unusual agreement between the NIH and the family of an African American woman, mother of five children, who had died in 1951 of aggressive cervical cancer, at the age of only thirty-one.[1] The woman was Henrietta Lacks, and the agreement concerned access to information about her genome derived from a cell line that biomedical researchers had long known as HeLa. Under the terms of the agreement, future researchers would have access to Lacks's genomic data only after a review that would include members of her family. An inanimate scientific tool, the HeLa cell line, acquired in this way a classic marker of personhood—the right to give informed consent, via living representatives of the person those cells came from, before the information it contained could be used for further research.

The agreement and the story behind it are unique in the annals of medical research. Just as David Leon Riley (see chapter 6) might have remained another lost soul among our incarcerated masses if a Stanford law clinic had not turned him into a constitutional case, so Henrietta Lacks might have lived as a footnote in medical history if Rebecca Skloot, a twenty-first century science writer, had not resurrected her story and revived her name.[2] Skloot first heard of Henrietta when, as a sixteen-year-old, she learned about HeLa cells in a lesson on cell division in a community college biology class. Her curiosity about the woman grew as she worked her way to an undergraduate biology degree, finding HeLa cells everywhere, in scientific publications and even in her own lab work. Ten years of single-minded research led to Skloot's phenomenally successful 2010 book, *The Immortal Life of Henrietta Lacks*: an account of a poor, uneducated, but remarkably resilient black woman who died of untreatable cancer, but whose cells, removed without her or her family's knowledge or consent, lived on as one of the most useful tools in the burgeoning field of biomedical research. In sixty years, HeLa cells had by some estimates generated billions of dollars in profits and more than sixty thousand scientific articles.[3]

Henrietta Lacks's story touched a national nerve with its blend of race, bioethics, economic and social inequality, and a young mother's untimely death. Most people did not know that tissues extracted from a person's body could be immortalized for scientific use long after the person had died, or that such use did not always require the donor's approval or consent. As presented by Skloot, the case cried out for acknowledgment that an injustice had been done, maybe even that some reparation should be made. Some people, after all, had grown rich from research using HeLa cells, while Henrietta's family remained poor and unable to afford even basic medical care. But it was another turn in the story, the prospect that Henrietta's cells would acquire yet a third lease on life, as a fully sequenced genome, that moved the NIH to take precautionary action.

Cells are material things. They can be grown in a dish, fed with nutrients, exposed to toxins, made to glow, and photographed and counted by means of the sophisticated instruments available to modern science. But in the era of genomic medicine cells are, as we have seen, more importantly also libraries of information. Each cell, of course, contains the genome, or the full genetic code, of the living entity it came from, whether a person, a bacterium, or a plant. That information can be used for diagnostic purposes, to study links between genes and undesirable conditions in the organism they came from. In human beings, for example, genomic information provides not only markers of physical or mental characteristics, such as eye color or mathematical ability, but also a basis for predicting whether a person is susceptible to one or another hereditary disease. Since genes are passed on from generation to generation, information contained in one person's genome provides information not just about that person but about family members as well. The sister, mother, or aunt of a woman diagnosed as having a mutant BRCA gene, which greatly increases the risk of breast and ovarian cancer, may carry the same mutation and be susceptible to the same risk. Genomic information, in this sense, can never be wholly personal: it is also information about a person's family, clan, tribe, or ethnic community.

In March 2013, Lars Steinmetz and his team of scientists at the European Molecular Biology Laboratory (EMBL) in Heidelberg, one of Germany's preeminent centers for biomedical research, sequenced the genome of the HeLa cell. Given the issues of property and consent already raised by Skloot's book, it was unlikely that American researchers would have walked into the swirling bioethical debate around the HeLa cells with eyes shut. Yet Steinmetz's team in Europe did just that. Seemingly unaware that their work might arouse controversy, they published the HeLa genome sequence in an online medical journal, *G3: Genes, Genomes and Genetics*. Dismayed and outraged that they were not consulted, and backed by Skloot's powerful advocacy, the Lacks family requested immediate retraction of the *G3* article. The EMBL researchers readily complied, but an American team based at the University of Washington was preparing at the same time to publish yet more detailed data on the HeLa genome.[4] The NIH director Francis Collins saw a threat to science unless leaders of the research enterprise found a comprehensive solution. Unsurprisingly, he invoked the popular metaphor of science running ahead of law and policy: "This latest HeLa situation really shows us that our policy is lagging years and maybe decades behind the science. It's time to catch up."[5] In practice, of course, scientists had long been making informal ethical judgments about appropriate and inappropriate forms of behavior in using human biological materials, and the NIH in this case simply continued that practice, albeit more publicly. The result was the historic deal granting future researchers access to the sequence of the HeLa genome, but only after review with the Lacks family present at the table.

The precise circumstances of Henrietta Lacks's life and death, and the extraordinary growth and vitality of the HeLa cell line, will never be duplicated. The case nevertheless calls attention to the fraught questions of ownership and control that arise when science and technology move into newly lucrative areas of production. Where, to start with, should lines be drawn between property and personhood? Further, which discoveries and inventions should society reward and how, and when, if at all, should profits be shared between inventors and others involved in the production of new knowledge? The answers lie partly in the territory of intellectual property rights—a legal domain as opaque, as technical, and as inconclusively charted as the zones of invention that it seeks to regulate. Partly, too, answers can be found in evolving notions of what counts as property when science and technology create entities that cross the lines between living and nonliving, human and nonhuman.

THE REWARDS OF INVENTION

Modern intellectual property rights are often traced back to the fifteenth century, though kings and governments encouraged and rewarded inventors from the beginnings of human history.

Sometime around the early modern period, a principle began to emerge that those who invent something of value to the state (and, later, of public value) should be granted exclusive rights to the fruits of their inventions—though not necessarily for all time. The founders of the American Republic took that principle so much to heart that they wrote it into the Constitution. Article 1, Section 8, provides, "The Congress shall have power . . . To promote the Progress of Science and useful Arts, by securing for limited Times to Authors and Inventors the exclusive Right to their respective Writings and Discoveries." To implement that right, Congress enacted the nation's first patent law in 1790, to be quickly superseded by a less administratively burdensome version in 1793.

Thomas Jefferson, who as secretary of state served as the nation's first patent examiner, found the notion of a monopoly on intellectual property particularly irksome. Exclusive control over ideas and inventions did not sit well with his Enlightenment imagination, which perceived the free sharing of knowledge and ideas as a cornerstone of liberal democracy. In a much cited letter of 1813 to the Boston mill owner Isaac McPherson, Jefferson compared an idea to fire, which can be passed from one to another without any loss of illumination: "Its peculiar character, too, is that no one possesses the less, because every other possesses the whole of it. He who receives an idea from me, receives instruction himself without lessening mine; as he who lights his taper at mine, receives light without darkening me."[6] In contemporary economic terms, an idea is a nonrivalrous good. Use by one person does not limit or diminish its utility to others. Society, Jefferson observed in the same letter, "may give an exclusive right to the profits arising from them as an encouragement to men to pursue ideas" of use to society, though, generally speaking, unlimited "monopolies produce more embarrasment [*sic*] than advantage to society."

Besides time limits on monopolistic rights, which Benjamin Franklin and others also favored, U.S. intellectual property law included further provisions to ensure that patents would bring more benefits than costs. Most important was a restriction on the kinds of things that could be patented. Technically known as "patentable subject matter," this list included under the 1793 Patent Act "any new and useful art, machine, manufacture or composition of matter," as well as new and useful improvements of any of those items.[7] Jefferson himself seems to have introduced the term "composition of matter," which had not previously appeared in U.S. patent law.[8] By definition, the list excluded things that existed in nature, and hence could not be regarded as products of human ingenuity, as well as things already known, hence not new or not useful. In his letter to McPherson, Jefferson held forth at length on inventions that would not be patentable, because they represented only a change in application, material, or form from something already patented. Such minor modifications would not fulfill the law's central purpose of encouraging and rewarding inventiveness of true value to society.

The procedures for granting and contesting patents have changed beyond recognition since Jefferson's day. An important institutional reform came in 1982 with the formation of the new Court of Appeals for the Federal Circuit (CAFC), a specialized, twelve-judge bench that hears all patent appeals. The subject matter clause of the law, however, remains remarkably close to the eighteenth-century original, with only the word "process" substituting for the older "art":

Whoever invents or discovers any new and useful process, machine, manufacture, or composition of matter, or any new and useful improvement thereof, may obtain a patent therefor, subject to the conditions and requirements of this title.[9]

Additionally, the law specifies that the invention must be nonobvious, that is, not immediately derivable from existing inventions by someone familiar with the state of the art in the relevant field. The meanings of all three key legal terms—"new," "useful," "nonobvious"—as well as the terms defining patentable subject matter have been continually reinterpreted to fit changing conditions in science and industry. Particularly significant from the standpoint of democratic values are the developments related to biotechnology. As Henrietta Lacks's story illustrates, cases involving ownership of any aspect of life raise troubling ethical questions about the appropriate balance between the demands of free scientific inquiry, technological inventiveness, and the sanctity and integrity of human life.

BODIES, CELLS, AND SELVES

The NIH's historic compromise with the Lacks family, a deal that allowed HeLa cells to keep circulating in science labs but controlled access to the informational content of those cells, was born out of unique circumstances. It reopened painful reminders of patterns of domination and discrimination in biomedical research that the nation's top scientists wished to put behind them once and for all. The NIH's leadership was determined to treat the decision as answerable to ethics, not commerce. No money changed hands in the NIH-Lacks deal, although some family members did raise the question of compensation. But all agreed that the Lacks case should be seen as sui generis. It was not generalizable. Indeed, as NIH's deputy director Hudson emphasized, "It's not going to be a precedent."[10]

The HeLa settlement thus provided no general answers to a more pervasive question that was left open in chapter 5. To whom do extracted biological materials belong, and who is entitled to profit from the medically useful derivatives that might result from their use? Medical processes of all sorts extract cells and tissues from patients' bodies, such as blood, urine, saliva, or tissues from surgically removed organs. Biobanks around the United States store hundreds of millions of such biospecimens, collected long before genetic analysis was easily available, and thus before the breaching of the line between biology and data or information. There were some early boundary-probing questions about ownership, but these were resolved in state courts, piecemeal, without adding up to coherent public policy. As a result, it remains murky how the law should govern intellectual property rights in information extracted from human biological specimens.

An early test arose in California in the 1970s. John Moore, a Seattle businessman, was only thirty-one when, in 1976, he headed to the University of California at Los Angeles (UCLA) to be treated for hairy cell leukemia, a rare cancer that was bloating his spleen and nearly killing him. The UCLA physicians, led by Dr. David Golde, removed his spleen and saved his life. Unbeknownst to Moore, however, they continued to do research on the tissues removed from his body, which they developed into a patented cell line named Mo. Moore returned for periodic follow-up tests over the next seven years, but he became suspicious when the UCLA researchers suddenly began insisting that he sign a consent form relinquishing all rights to products derived from his tissues. Only then did he learn the extent of the UCLA researchers' commercial interests based on the Mo cell line and other patents.[11] Moore eventually sued the university and his doctors on the unprecedented ground that, in removing and working with his diseased spleen, they had stolen his property for their own use.

California courts had never been asked to consider the question of profound moral significance that Moore's claim raised. What is the relationship between our bodies and our selves? Are our bodies like physical property, so that we can completely control what becomes of them? Or are we, as autonomous agents, entitled only to give consent to physical intrusions, such as medical procedures, but not to direct what should be done with excised tissues or other remains? The Supreme Court of California concluded in 1990 that John Moore did not own his cancerous cells in any ordinary sense of the term and that their removal and use in biomedical research was therefore not unlawful.[12] Where Dr. Golde and the university had erred was in not disclosing their commercial interests in advance, and thus not obtaining Moore's consent to serve as a de facto research subject. After seven years of litigation, Moore received token compensation for that oversight. He became an advocate for patient's rights for the next twenty years, until he died of cancer in 2001, at the age of fifty-six.

Another case that underscored the confused lines between personal autonomy, bodily integrity, doctor-patient relations, and commercially profitable invention brewed in the 1980s at Washington University (WU) in St. Louis, Missouri.[13] Dr. William Catalona, a well-known prostate surgeon and longtime employee of WU, had treated thousands of patients who had consented over the years to serve as research participants and to let him store their tissue samples for studies in which he was the principal investigator. In 2001, Catalona approached a biotech company with some of his samples in exchange for help in evaluating a genetic test for prostate cancer. The university's technology management office got wind of his plan, saw a potential windfall for the university, and tried to negotiate a deal that would bring more economic benefit to WU. Possibly frustrated at the interference with his plans, Catalona decided in 2003 to leave WU for Northwestern University and began preparing the ground for taking a part of the tissue collection with him. He sent a letter to some sixty thousand WU research participants, requesting them to sign a release form containing the following language:

> I have donated a tissue and/or blood sample for Dr. William J. Catalona's research studies. Please release all of my samples to Dr. Catalona at Northwestern University upon his request. I have entrusted these samples to Dr. Catalona to be used only at his direction and with his express consent for research projects.

Catalona received six thousand releases, but once again the university stepped in, this time suing him to block the transfer and to gain ownership of the stored materials.

After some five years of legal wrangling, the federal Court of Appeals for the Eighth Circuit ruled in favor of the university.[14] The main question, as the court saw it, was whether persons who knowingly donate their biological samples to a research institution for the purpose of medical research retain any rights to transfer ownership to a third party. The court's answer was no. Under the consent forms they had signed when they entered a study, the court concluded, participants had made a voluntary gift to WU. The only rights they still retained were to revoke their consent to further research and possibly to have their samples destroyed. Redirecting the samples to Dr. Catalona was simply not an option.

Unlike some countries, the United States has not enacted any comprehensive national policies regarding the ownership of biospecimens.[15] Nonetheless, the cluster of state court decisions addressing tissue ownership points to a priority scale in which research leading to drug discovery and commercial benefits is valued above individual interests in protecting biological samples. Although one state's legal decisions are not formally binding on any other state, *Moore* influenced the Eighth Circuit's thinking in the Catalona case. *Moore, Catalona*, and the handful of related cases left research participants with few rights to their physical specimens, let alone to information and products derived from them. Material drawn from diseased bodies thus became mere matter, and not matter subject to the individual patient's control. As we see below, this way of drawing the lines drew justification from a fear of hindering technological progress—a fear that looms large across the spectrum of judicial decisionmaking on biology and intellectual property, even when courts seek to rein in perceived excesses in turning human selves into marketable commodities.

REINVENTING LIFE FOR PATENTS

What did the framers of the early patent laws have in mind when they listed as patentable subject matter "any new and useful art, machine, manufacture or composition of matter"? The examples they used offer some guidance. In his letter to McPherson, Thomas Jefferson illustrated his argument with the homeliest objects—hats, shoes, combs, buckets, milling and agricultural implements—all utilitarian and all thoroughly inanimate. Patent practices in the first years after the law's enactment offer additional insights into what the early legislators saw as appropriate subject matter. The first U.S. patent was issued on July 31, 1790, to Samuel Hopkins for a new apparatus and process for making the potassium-based industrial compounds potash and pearl ash. Only two other patents were granted in that first year, one for a new way of making candles and the other to Oliver Evans, a prolific inventor and an adversary of Isaac McPherson's, for an automated flour mill. It was not in doubt that these objects and processes were patentable; nor was their utility in question. Only their novelty needed to be ascertained. Early conflicts over the patent system centered, not surprisingly, on costs and delays, which inventors found aggravating or intolerable, in establishing those claims of newness.

By the last quarter of the twentieth century, however, industrial production had moved beyond its traditional involvement with machines, chemicals, steel, and other hardware of daily use to encompass living biological entities. Designs on nature became the centerpiece of what some saw as a second industrial revolution, eventually also embracing advances in nanotechnology, cognitive sciences, and information technologies. New kinds of valuable commodities came into being, straining older conceptions of what counts as patentable subject matter and indeed as property more generally. Legal judgments appeared to settle some questions, but new ones arose as inventors' claims bumped up against deep-seated cultural expectations that some aspects of life should not be subject to claims of private ownership.

Patents on Life

In the early 1970s, Ananda M. Chakrabarty, an American biochemist born and educated in West Bengal, India, joined the General Electric (GE) Research Center in Schenectady, New York, where he set to work designing a bacterium capable of breaking down a mix of hydrocarbons in oil. Chakrabarty discovered that the genes allowing bacteria to degrade oil reside not in the bacterial chromosome but in rings of DNA known as plasmids that can be transferred between organisms. Tinkering with these bacterial plasmids eventually paid off. In time, Chakrabarty created a new oil-degrading variety of the *Pseudomonas* bacterium containing fused bits of DNA from the plasmids of four existing bacterial strains.

When his superiors at GE learned of his achievement, they advised him to do the normal thing in any industrial research laboratory: patent his invention, even though bacteria lay far outside GE's usual product lines. Chakrabarty told the historian Daniel Kevles that a drug company probably would have hesitated to patent a living organism, deterred by the "product of nature" exemption, which says that no patents can be issued on things found in nature. But Leo I. MaLossi, the GE patent lawyer assigned to the case, "was used to filing patent applications on items like refrigerators, plastics, jet engines, and nuclear power plants, and thought that if you invented something new and useful, you deserved a patent covering whatever claims about it you could legally make."[16] MaLossi plunged ahead, even though he had to persuade the U.S. Patent and Trademark Office (USPTO) and eventually the courts that, for purposes of patent law, a human-made bacterium was no different from any other item in GE's inventory of useful machines.

Chakrabarty's claim rode a roller coaster of uncertainty until it reached the Supreme Court in 1980. The USPTO initially denied his claim on the ground that living organisms could not be patented under existing law. The case then moved up the ladder to the Court of Customs and Patent Appeals, predecessor of today's Federal Circuit, where it received two separate hearings—before and after a Supreme Court decision on the patentability of mathematical algorithms. Both times, the appellate court reversed the USPTO and upheld Chakrabarty's claim. The patent and trademark commissioner then appealed to the Supreme Court, where the justices ultimately held in a 5–4 decision that nothing in the law barred the grant of a patent for Chakrabarty's bug. The difference between living and nonliving organisms, the majority concluded in *Diamond v. Chakrabarty*,[17] was irrelevant to the inventors' product patent claim. Quoting a congressional report of 1952, the Court endorsed a virtually limitless interpretation of the subject matter clause: patents may be awarded for "anything under the sun that is made by man."

The majority's disinclination to limit patentable subject matter to nonliving things derived strong support from arguments made by amici curiae, or friends of the court. One submission that proved especially influential was an amicus brief filed by Genentech, a biotech start-up founded by Herbert Boyer, codiscoverer of the technique of gene splicing or recombinant DNA (rDNA). Research funded by Genentech had produced the first lab-created human insulin in 1978. Scientists used rDNA techniques to synthesize the insulin-producing gene and inserted it into *E. coli* bacteria, which then acted as living "factories" for making insulin. The process offered an alternative to the earlier method of extracting insulin from the pancreatic glands of cows and pigs. This process shift was expected to make insulin both cheaper and more abundant.

Genentech's brief to the Supreme Court sought to downplay any resemblance between Chakrabarty's bacterium and the kinds of life whose sanctity the law rightfully seeks to protect. In effect, the company argued, the requested patent was not even for a living entity but for plasmids, inanimate rings of DNA, inside a bacterial body. Plasmids, surely, were "dead chemicals" quite properly covered by the phrase "composition of matter." Moreover, Genentech noted, the USPTO seemed quite willing to grant patents on mixes of plasmids and straw. What was so different about a plasmid inside a bacterium? The brief posed the question rhetorically: "Can it be said that Congress intended patents on living organisms inside inanimate bits of straw but prohibited them in the case of inanimate bits of chemical inside microorganisms, or are we beginning to draw distinctions that border on the silly?"[18]

That appeal to common sense aligned well with the prevailing culture of judicial decision making. Common-law judges are conditioned to buy into incremental arguments, and they do not like using judicial authority to anticipate or intervene in speculative, faraway futures.[19] As courts see it, that function of foresight-based policymaking belongs more properly to legislators. Law, by contrast, looks backward at precedent, and forward chiefly to guard against immediate, easily imagined harms. Predictably, therefore, the *Chakrabarty* majority found Genentech's arguments more persuasive than the more distant, dystopian, and, in their view, political considerations urged on them by Jeremy Rifkin, the era's most vocal critic of uncontrolled developments in genetic engineering. Rifkin's organization, the People's Business Commission (PBC), warned in its amicus brief that patenting anything living would inevitably lead to more extensive commercialization of life: "if patents are granted on microorganisms there is no scientific or legally viable definition of 'life' that will preclude extending patents to higher forms of life."[20] The Court rejected this vision of a slippery slope, insisting from the first line of its opinion that its sole, narrow concern was "to determine whether a live, human-made microorganism is patentable subject matter under 35 U.S.C. § 101."

With the clarity of hindsight, Rifkin's position looks prescient in several respects. Just as the PBC had predicted, *Chakrabarty* opened the door to patenting higher forms of life. Once the line between life and nonlife was ruled irrelevant, no one could see good reason to deny patents on oysters, mice, or larger mammals. To be sure, the USPTO moved forward cautiously. It waited until 1988 to grant a patent on the oncomouse, a lab animal genetically engineered by Harvard University researchers to be predisposed to cancer and therefore of great potential value in testing cancer drugs. Responding to a public outcry, and fearful of adverse action by Congress, the government then adopted a voluntary five-year moratorium on patenting animals, resuming only in 1993.[21] Since then, the United States has issued hundreds of patents on genetically engineered animals, though not all countries have followed suit and none with equally unbridled enthusiasm.

The European Patent Office (EPO) also allows animal patents, but subject to the constraint in European law that patents should not contravene *ordre public*, a French term that literally means "public order" but is often translated in the patent context as "morality."[22] The morality clause requires the EPO to weigh the claimed benefits of the invention against such negative consequences as the suffering of the animal, adverse effects on the environment, or public moral unease. Two patent applications, one granted and the other denied, illustrate the impact of such balancing. The first concerns the European patent for the Harvard oncomouse, which the EPO approved in 2004 after repeated evaluations, but it limited the patent to mice and excluded any other nonhuman species. The second case concerned a hairless mouse produced by the Upjohn Company to test products for hair restoration and wool growth. Whereas in the case of the oncomouse there were potential medical benefits, here the EPO found no such benefits to offset the impacts on the mouse, and hence it refused to grant a patent.

The Canadian Supreme Court in 2002 chose yet another path, denying a patent for the Harvard oncomouse and at the same time ruling broadly against patents on all higher animals. That decision is particularly interesting because Canada uses virtually the same language as the United States to define patentable subject matter (Canada retains the word "art" along with "process, machine, manufacture or composition of matter"). Yet, a majority of the Canadian Supreme Court refused to go along with the U.S. view that animals should be treated simply as just another "composition of matter." A mouse, even an oncomouse, has qualities and characteristics that lift it above its genetic composition, the court concluded. Animals, according to this view, are not simply engineering scripts applied to what Genentech termed dead chemicals. Instead: "The fact that it has this predisposition to cancer that makes it valuable to humans does not mean that the mouse, along with other animal life forms, can be defined solely with reference to the genetic matter of which it is composed."[23]

The Canadian court did not address what divides a higher animal that cannot be patented from a lower organism that can be; nor has such a criterion been articulated anywhere else. We know only that, in Canada, bacteria are patentable and mice are not. Even in the more permissive U.S. regime, human beings stand outside the reach of patents, although just how far outside has yet to be defined. The issue is likely to become more critical as biological research produces new human-animal hybrids, or chimeras, consisting in part of human stem cells inserted into animals such as rodents. In 1997, Jeremy Rifkin teamed up with Stuart Newman, a developmental biologist at New York Medical College, to provoke the USPTO into drawing a bright line on the patentability of chimeras. They applied for a patent on a hypothetical cross between a human and a chimpanzee—dubbed a humanzee[24]—based on several proposed methods of constructing such a hybrid. The USPTO turned down their application, initially in 1999 and again in 2005, the second time acting on legislative authority. Congress in 2004 added an amendment to an appropriations bill banning the use of federal funds "to issue patents on claims directed to or encompassing a human organism." While the exact meaning of the words "directed to or encompassing" remains unclear, the USPTO said that the humanzee described by Newman might be close enough to a human to fall within the scope of the congressional prohibition.[25]

Backtracking on Genes

As we have seen, it is a settled principle of patent law that things existing in nature cannot be patented. Merely finding something that is already there, so the logic goes, is not inventive and should not entitle the finder to any special rights of ownership. Besides, finding one previously unrecorded tree or one new kind of gemstone should not confer a right to own all trees or all gemstones of those kinds. Biotechnology, however, tested the limits of this doctrine as techniques became available to isolate and purify objects found in nature, but only in mixed or impure forms, as well as to synthesize complex entities from more basic materials. The processes of purification and synthesis were unquestionably patent-worthy, but inventors and start-up companies argued that isolated strings of DNA that code for valuable proteins were not "products of nature" and should also be patentable. To support that view, courts and companies cited a 1911 decision by the influential Judge Learned Hand, upholding a patent on adrenaline purified from animal glands.[26] By the early 1990s, the USPTO was routinely granting patents on isolated as well as synthesized DNA sequences, including human genes. Biotechnology companies derived huge profits from these patents, which gave them exclusive rights to develop diagnostic tests and other products based on isolated genes. Few expected the clock to be turned back on that policy.

At the American Civil Liberties Union (ACLU), Tania Simoncelli saw things differently. Much as the GE lawyer MaLossi had plowed ahead where lawyers for pharmaceutical companies might have feared to tread, so Simoncelli, who lacked formal legal training, was not deterred by presuppositions about the fixity of the law. She had joined the ACLU in New York in 2003 as the organization's first science adviser. Simoncelli brought to her new (and as yet poorly defined) job years of immersion in public interest activism, dating back to her undergraduate years at Cornell, as well as specific interests and competence in policy issues related to biotechnology. She had watched the growing commodification and commercialization of living matter with dismay, ever since she read *Diamond v. Chakrabarty* in my science and law class as part of her major in biology and society. She became convinced that patents on human genes were wrong and that most sensible people, their minds unclouded by the intricacies of patent law, would share her view. At the ACLU, she got to test her intuitions and lined up powerful allies to attack a seemingly unshakable legal principle.

Chris Hansen, then senior staff counsel at ACLU, could not believe it when Simoncelli first told him that companies were patenting human genes. "That's ridiculous!" he recalls saying. "Who can we sue?"[27] The answer it turned out was not so simple. Hansen was an experienced litigator but not a patent attorney. The lawsuit he mounted in partnership with Simoncelli took seven years to reach a victorious end. The first challenge, as Hansen immediately saw, was to identify a promising defendant: somebody to sue. But it was no easy matter to line up plaintiffs who would be recognized and admitted by the courts, or to enroll the scientific experts who would help build the technical arguments in the case. Each task demanded years of preparation and effort, and the outcome hinged on luck at key steps along the way.

The defendant the ACLU team ultimately set its sights on was Myriad Genetics, a Utah-based company that held the patents on the cancer-causing BRCA genes and also marketed lucrative tests, trade-named BRCAnalysis©, to detect the presence of those genes. More than most successful patent holders, Myriad was ruthless in protecting its market. The company had taken firm action against scientists and clinics that it saw as encroaching on its turf by conducting BRCA tests of their own. One target was Dr. Harry Ostrer, a medical geneticist at the Albert Einstein College of Medicine, in New York. Way back in 1998, Myriad had sent Ostrer a demand for royalties under its patents if Ostrer tried to offer breast cancer diagnostic testing at his clinic. Ostrer retained the letter, which, years later, gave him a documented basis for claiming that Myriad's patents were preventing him from conducting BRCA testing. With that claim of real injury, Ostrer proved to be the one plaintiff who survived all attempts to exclude him. Simoncelli and Hansen also lined up heavyweight scientific support from the academic community, most notably from Eric S. Lander, founding director of the Harvard-MIT Broad Institute and cochair of President Barack Obama's council of scientific advisers.

Though many factors aligned in the ACLU's favor, the effort to knock down a long-established USPTO policy, potentially affecting thousands of patents and the enormously successful U.S. pharmaceutical industry, struck many as quixotic, even frivolous. The outcome remained in doubt until the end, and, as in the *Chakrabarty* case, there were big ups and downs along the way. Luck intervened on the ACLU's side when the case was randomly assigned to Judge Robert W. Sweet in the federal District Court for the Southern District of New York. Born in 1922 and appointed to the federal bench by President Jimmy Carter, Judge Sweet had retired to senior status in 1991 but was still actively hearing cases. His law clerk at the time was Herman H. Yue, a Berkeley-trained geneticist whose technical knowledge proved invaluable to the judge's consideration of the case.

Judge Sweet's decision in 2010 to invalidate Myriad's BRCA patents caught everyone by surprise—not least Simoncelli and Hansen, who had endured a drumbeat of commentary saying that their case could not be won. Yet the principle on which the decision rested was nothing other than the tried and true "product of nature" doctrine. To be eligible for a patent, the court said, the product had to meet the subject matter test, and this the BRCA genes failed to do because they were not "markedly different" from genes in the human body, as required by the *Chakrabarty* decision. To the contrary, "isolated" DNA is identical with native DNA in the body both in its biological functions and in the information it encodes. Judge Sweet concluded, "Therefore, the patents at issue directed to 'isolated DNA' containing sequences found in nature are unsustainable as a matter of law and are deemed unpatentable subject matter under 35 U.S.C. § 101."[28]

It was a remarkable first-round victory, but two more rungs of the federal judicial ladder had to be climbed before the ACLU could declare that it had actually won its case. Myriad appealed the district court decision, and the next steps went less smoothly for the challengers. A three-judge panel of the CAFC, regarded by critics since its foundation as probusiness and propatent,* ruled twice in Myriad's favor, each time by a 2–1 majority, once before and once after the Supreme Court asked for reconsideration. Both times, the majority opinion concluded that isolated genes met the criteria of patentability; both times, too, the majority expressed grave hesitation about undermining a flourishing industry by dismantling one of its most dependable architectural supports.

In a curious twist, the Justice Department filed an amicus brief supporting ACLU's claim that Myriad's gene patents were invalid. Arguing against the USPTO, the government tried to steer a middle path that would give Myriad much of what it wanted, but not the ultimate prize of owning the isolated BRCA1 and BRCA2 genes. The brief distinguished between *isolated* DNA and *complementary* DNA (cDNA), the latter synthesized from a messenger RNA (mRNA) using an enzyme called reverse transcriptase. To support this assertion of difference, the government's brief turned to fantasy, arguing that an isolated gene could not be patented, because it would fail the "magic microscope" test: "[I]f an imaginary microscope could focus in on the claimed DNA molecule as it exists in the human body, the claim covers ineligible subject matter."[29] Human-made cDNA, however, would not be visible through such a microscope because, unlike natural DNA sequences, cDNA consists of only the coding sections of genes (exons) and not the intervening

---

*The CAFC was established in part because patent rulings were felt to require specialized expertise and a consistent approach to patent policy, but since its formation many have perceived the CAFC's judgments as lacking critical depth and even as being captive to the patent bar. See, e.g., David Pekarek-Krohn and Emerson H. Tiller, "Federal Circuit Patent Precedent: An Empirical Study of Institutional Authority and IP Ideology," Northwestern University School of Law Scholarly Commons, Faculty Working Papers, 2010.

noncoding sections (introns) that make up the entire strand as it exists in nature. Therefore, although cDNA is functionally like the corresponding DNA in its ability to express proteins, one would not find exactly such a complementary sequence if one took a microscopic look inside the human body.

The government's ingenious analogy did not persuade the CAFC. Looking at a molecule through a microscope struck the court as "worlds apart from possessing an isolated DNA molecule that is in hand and usable." Looking was simply discovery, the job of science and of satisfying one's curiosity. Isolating a gene, by contrast, manipulates matter, produces something useful, something not previously found in nature, and hence is precisely the kind of technological inventiveness the Patent Act seeks to encourage. But Eric Lander's brief to the Supreme Court dealt a significant setback to the CAFC's position. From his position of the highest scientific as well as policy authority, Lander informed the Supreme Court that the CAFC was wrong to think that isolated DNA does not occur inside the body. On the contrary, "[i]t has been well established for over 30 years that isolated DNA fragments of human chromosomes routinely occur in the human body. Moreover, these isolated DNA fragments span the entire human genome, including the BRCA1 and BRCA2 genes."[30]

It was now up to the Supreme Court to make the final ruling. The question put to the Court was brief and to the point: "Are human genes patentable?" In response, the Court unanimously ratified the Justice Department's middle-of-the-road position. Isolated genes are the same as DNA sequences found in the body; therefore, as products of nature they cannot be patented. On the other hand, cDNAs consisting of only the exons from the naturally occurring molecule are not products of nature and thus can be patented. After years of litigation, contradictory rulings, and persistent uncertainty, it all seemed suddenly so simple. Even Justice Antonin Scalia curbed his fondness for verbal pyrotechnics and limited himself to a single paragraph concurring with his colleagues' judgment.

RIGHTS ACROSS BORDERS

Biological information and the capacity to turn it to medical and commercial benefit are distributed today among a growing variety of actors. Each controls an essential piece of the platform on which new inventions must ultimately be built. Individuals provide biological materials and family histories; clinics gather specimens that go into biobanks; biotech companies manipulate cells and genes to make purified, isolated, and synthetic products; and pharmaceutical companies conduct research and development to move a test or a drug from concept to market. Even when a single person is credited with a breakthrough discovery, like the American biochemist Kary Mullis, who won the 1993 Nobel Prize in Chemistry for his improvement of polymerase chain reaction (PCR), such work builds on economic, social, and legal infrastructures without which the award-winning discoveries would never have produced valuable goods.[31]

The idea of the lone ingenious inventor that animated eighteenth-century Western intellectual property law seems too limited to handle the complex claims of ownership that grow out of the twenty-first century's highly distributed knowledge systems. The lack of fit becomes most evident when intellectual property claims cross national boundaries. Several attempts have been made to regulate those borderlands through international treaties and agreements, but it remains a poorly controlled space, where both piracy and unjust enrichment pose serious problems.

Indigenous Knowledge

One set of difficulties concerns the relationship of indigenous knowledge to discoveries made and marketed by modern biomedicine. In colonial times, military conquerors and missionaries freely appropriated the knowledge and practices of local people using traditional, plant-based medicines against various illnesses. That knowledge was transported back to the centers of imperialism, where it provided the basis for a nascent pharmaceutical industry. Already in the 1630s, for example, Jesuit missionaries sent cinchona bark from Ecuador and Peru back to Europe, where it was used to treat shivering induced by malarial fever. In time, cinchona became the source of a lifesaving drug, quinine. Value was extracted, literally and figuratively, from indigenous knowledge and resources, but with no payback.

To many observers, the problems of today appear to continue historical patterns of subjugation and unjust enrichment. Although native healers possessed resources of great value to humanity, the ideas of novelty, utility, and nonobviousness that define Western intellectual property law do little to protect the collective knowledge of preindustrial societies—where knowledge is communally held and health benefits accrue to the population as a whole. The 1992 Convention on Biological Diversity (CBD) sought to rectify this imbalance by providing that indigenous people should receive adequate returns for the use and exploitation of their knowledge and resources. Benefit sharing is one of the CBD's principal objectives, along with the conservation of biodiversity and sustainable development of its components. Article 8(j) directs parties to "respect, preserve and maintain knowledge, innovations and practices of indigenous and local communities" and to "encourage the equitable sharing of the benefits arising from the utilization of such knowledge, innovations and practices."

These provisions aimed to encourage "bioprospecting," or mining plant and microbial genetic resources to generate revenues that would encourage biodiversity protection: in other words, "selling biodiversity to save it." The CBD also laid the basis for technology transfer by encouraging wealthy research institutions in the global North to locate less well-endowed partners in the global South who would agree to share local knowledge in exchange for advanced technology and know-how. Such agreements often misjudged the capacity of local partners to make binding agreements and gave rise to noisy arguments about who actually speaks for nature in such communities.[32] Charges flew that the CBD offered an open invitation to biopiracy. Nonetheless, some analysts saw positive steps in institutional learning in the first two decades of implementing the CBD.[33]

Generic Drugs

Another set of problems grows out of contemporary socioeconomic inequalities that intellectual property regimes tend to hold in place. Rich nations can produce, access, and pay for new technological products and services much more easily than poor nations can. As a result, there are many incentives in poorer nations to circumvent the monopolies created by Western patent regimes, either through piracy or by infringing on patents. A practice with immense consequences for global public health is the production of generic versions of patented drugs that can be marketed at a small fraction of their original cost. Before 1994, many developing countries had excluded pharmaceutical drugs from patent protection on the ground that they are necessary for preserving life and health. That exemption allowed countries like India, Brazil, and Argentina to develop a large capacity for producing generic drugs, chemically identical with or closely similar to their patent-protected variants in the global North but much cheaper. Generic drugs, according to northern manufacturers, left them doing the hard work of invention, and shouldering the risks of failure, while returning profits to companies that did not innovate but were simple imitators.

In 1994, these arguments came to a head at the international trade discussions known as the Uruguay Round, leading to a modification in trade law. To enjoy the benefit of lower tariff barriers under the General Agreement on Tariffs and Trade (GATT), signatories were required, under the Agreement on Trade-Related Aspects of Intellectual Property Rights (TRIPS), to adopt intellectual property laws offering copyright and patent protections similar to those then obtaining in developed countries. The law changed, but it failed to pacify critics, who saw TRIPS as one more measure to transfer wealth from poorer to richer countries. Under pressure to reform the trading system generally, World Trade Organization members agreed, in Doha in 2001, to loosen some of the pressures on less developed countries. A special declaration on TRIPS recognized that countries facing public health crises, such as HIV/AIDS, malaria, or tuberculosis, might need to break with TRIPS requirements in order to provide their citizens access to essential medicines. Countries may bypass patent protection under such circumstances through compulsory licensing to lower the price of the necessary drugs. Ironically, an early test for the revised provision came in the developed North rather than the developing South. In the wake of 9/11 and the subsequent anthrax attacks in America, the German drug company Bayer was forced to drastically lower the price of Cipro, a patented antibiotic remedy against anthrax, under threat of compulsory licensing by the United States.

Another signal of the instability of the current global patent regime for generic drugs came with a lawsuit in India involving the anticancer drug Gleevec, manufactured by the Swiss pharmaceutical giant Novartis. Indian firms began making and marketing the generic version of Gleevec, a crystalline form of the compound imatinib, before TRIPS, at a time when India did not yet authorize patents on pharmaceuticals. Later, after India passed a national law implementing TRIPS, Novartis filed for a patent on its version of the drug, which would have raised the price ten times higher than the available generic, thereby taking it out of reach for most Indian consumers. In 2013, the Indian Supreme Court held that Gleevec did not meet section 3(d) of the new Patents Act, which guards against patent renewals based on minor improvements that do not confer added therapeutic benefits—a practice commonly known as "evergreening."[34] In the court's opinion, Novartis had failed to show that the particular form of Gleevec it sought to patent was any more efficacious than cheaper off-patent versions already on the market.[35] It was merely evergreening its older product.

The Gleevec case took shape during a period when the Indian government was assimilating its existing patent protections to TRIPS. Novartis had applied for a patent on Gleevec twice, with somewhat different product specifications, at the beginning and end of this transition in the law. These exact circumstances are unlikely to be repeated for many other drugs. Yet the Indian Supreme Court went out of its way to underscore the nonneutrality of patent law, the close connection between political values and intellectual property protection, and the consequent mismatch between the ethical demands of developed and developing countries with respect to intellectual property. The court cited with apparent approval a 1957 report on patent reform authored by another judge:

> Justice Ayyangar observed that the provisions of the Patent law have to be designed, with special reference to the economic conditions of the country, the state of its scientific and technological advancement, its future needs and other relevant factors, and so as to minimize, if not to eliminate, the abuses to which a system of patent monopoly is capable of being put.[36]

The Novartis case settled the patentability of Gleevec in India, but it left open a larger ethical question. How much variance between patent systems is warranted if one accepts Justice Ayyangar's contention that intellectual property law is not value-free but articulates the political and economic preferences of particular nations or regions? An indication of possible ways forward came in September 2014, when Gilead Sciences, the California-based maker of a costly drug for treating hepatitis C, signed a license with seven Indian manufacturers of generic drugs to produce a stratified global pricing system.[37] Under the agreement, Gilead would sell its drug in India for ten dollars per pill, one-hundredth the price the company charges in the United States. In return, the Indian manufacturers would pay a licensing fee to Gilead but continue marketing their generic versions in poor countries where patients would never be able to afford the high-priced pills. This, however, was an ad hoc, private agreement between pharmaceutical companies in two countries without legal or precedential value for other drugs, firms, or patient populations.

CONCLUSION

Science continually turns the pages of the book of life, bringing new facts and inventive opportunities into view. The law's
function is to make sure that society reads and uses the content
of those pages in accordance with fundamental human values,
such as respect for life and care for its diversity and flourishing. Intellectual property law, whose main goal is to encourage
invention, is no exception. It is not required to reward anything and everything that discoverers and inventors dream
up. Indeed, the law can even turn back the pages if science
and technology seem to be flipping too quickly or heedlessly
ahead of widely shared values, as is especially likely to happen in a world where resource distribution remains extraordinarily unequal. In such cases, the assumption that invention is
always well aligned with the public good, at national or global
scales, can be revisited and critically questioned, with associated changes in policy and law.

The *ordre public* provision of European patent law reminds
us that moral limits can be placed on patentability. Canada's
refusal to allow a patent on the oncomouse set an upper bound
on the commodification of life. Even in the liberal United States,
Diamond v. Chakrabarty's reading of U.S. patent law as covering
"anything under the sun that is made by man" proved over
time to be too expansive by the standards of American ethics,
American law, and even American science. The NIH's agreement with the Lacks family and the Myriad gene patenting case
nicely illustrate the law's capacity for creative nonlinearity, to
the point even of reversing its own settled principles when society's values declare "this has gone too far."

Patent disputes also illustrate the power of individual action to transform taken-for-granted assumptions. Chakrabarty's attorneys successfully circumvented the "product of nature" doctrine by getting the Supreme Court to allow patents on life. By letting the Lacks family participate in controlling access to information from the HeLa genome, the NIH paid homage to Rebecca Skloot's single-minded crusade honoring the memory of the family's matriarch, Henrietta Lacks. The NIH's decision gave posthumous voice to a woman who had donated her body to research without her knowledge or consent. In the case of human gene patents, a bureaucratic agency, the U.S. Patent and Trademark Office, had interpreted the law to mean that isolated genes are eligible to be patented. What wider society thought about the issue was not tested until Tania Simoncelli and Chris Hansen launched their improbable campaign to overturn a policy that many experts thought was set in stone. When the matter at stake is life itself, it is reassuring that individual voices can tap into deeper currents and, in exceptional cases, bring about a realignment between lay values and esoteric law.

The justice implications of intellectual property law are most poignantly visible when patent claims set up disparities across borders, as when longstanding cultural knowledge in one locality provides the basis for blockbuster drugs and associated profits in another, or when monopoly pricing keeps essential medicines out of reach of patients who cannot afford them. Despite the attempt to globalize intellectual property protections through TRIPS, questions remain about the ethical implications of that move, not all of which were resolved through the compulsory licensing provision signed in Doha. Just as U.S. patent law had to make adjustments when faced with an excessive commodification of life, so the world community has some hard thinking left to do about the pharmaceutical industry's power to regulate life and death through strictly enforceable patents on essential drugs. Fortunately, drug companies, patient groups, activist physicians and lawyers, and the world trading regime are already involved in a conversation of profound significance for the ethics of global biomedicine.

Chapter 8

RECLAIMING THE FUTURE

Technology, as we have seen, is often defined in instrumental terms, as a means to a preordained end. But that way of thinking is too narrow to encompass the dynamic and multifaceted relations that modern human societies have forged with their all-pervasive tools and instruments. To begin with, the instrumental understanding of technology implies that human ends are well defined and static: we use tools because we have needs that must be satisfied, such as finding food or warding off bad weather. Yet employing tools for such mundane ends is not even a distinctively human action. Birds do it. The remarkable crows of New Caledonia use long sticks to dig wood-boring insects out of tree trunks or dead animals. In experiments, these crows have even learned to use multiple tools in succession and to bend wires to make better implements for grasping.

Jane Goodall, whose field studies in Tanzania's Gombe Stream National Park made her and her subjects famous, observed chimpanzees making and using tools for varied purposes. Leaves and moss make primitive sponges for drinking. Branches, twigs, and stones are used to pound and dip into termite nests, to smash and scoop out the insides of nuts or animal bones, and for sheer display. Elephants, too, are extraordinary tool users. They clean off branches to use as flyswatters or to dig for water, and they use stones to break things that get in their way. In captivity, elephants have been known to roll a cube into place and get up on it to grab a fruit placed tantalizingly high and out of reach.[1] With training, elephants can manipulate a loaded paintbrush accurately enough with the tips of their trunks to paint a standard picture over and over again.[2] What these impressive examples lack, though, is reflection on the nature of the task to be performed or any consequent change in the ends that the tool deployer took to be worth achieving from the outset.

Neither practicality nor predictability captures the evolving relations between human beings and their technologies. Human technological wizardry extends far beyond performances of repetitive tasks to serve simple, predetermined purposes. Artistry, imagination, and the desire to probe the unknown have long dominated the will to make and use technology. Turn the clock back some 35,000 years before the Common Era and enter the *Cave of Forgotten Dreams*, the famed German director Werner Herzog's 2010 exploration of some of the earliest figural art created by human hands. Here on the walls of the Chauvet cave in France, deep inside a cliff face above the old course of the Ardèche River, before it bent and cut its way through the magnificent opening of the Pont d'Arc, Ice Age artists drew and painted a stunning array of animals, some in motion, some artfully smudged like paintings on paper, some with sharply chiseled outlines for clearer definition. The art of the Chauvet cave is unique in its concentrated richness and technical mastery, but in that period art flowered more broadly, as ancient humans began to turn stone and horn and ivory into representations that had no practical uses. The makers imagined things that never were, such as a man with a lion's head and endless stylized figures of women, including one that inspired Picasso. Surveying all this abstraction and minimalism, the polished symbolism of a prehistoric age, a 2013 British Museum exhibition of Ice Age art declared that this period marked "the arrival of the modern mind."

If Ice Age cavemen (and perhaps equally women) used implements to paint and carve things that existed only in their minds, then how much more do we modern humans use technology to reimagine the most basic purposes of our existence. Who are we, what is life for, how should we live, what might the future hold? A giant leap beyond those first stirrings of aesthetic and symbolic sensibility came when technology began to serve not only individual purposes but collective ones as well. One could paint a bear or a bison or an ibex on a cave wall in solitude, though no doubt it mattered that others would see it. Today's technological imaginations can change the world more profoundly, for many more people, and often irrevocably. If synthetic biologists succeed in creating new organisms or if computer scientists figure out how to link thoughts across human brains, these inventions will alter our common understanding of where life begins and ends, what cognition means, and what is the human self. If neuroengineers figure out how to activate or suppress human memories through nanoscale implants in brains, then the mere availability of such a possibility will transform our present-day expectations concerning forgetfulness, and maybe forgiveness. If geoengineers manage solar radiation to keep the Earth from warming too much, the aerosols released into the atmosphere will whiten and brighten the skies, altering the experience of nature for everyone on the planet.

Technology, in short, is not merely about achieving ends that we already foresee but an open door to an uncharted, often uncertain future where current social understandings and practices may be fundamentally transformed. Uncertainty, moreover, can deter as much as it entices. The bright gleams of promise that invite human societies to invest in technology march hand in hand with darker misgivings about what could go wrong if the promises fail and the unexpected breakdown happens on a grand scale. Science fiction is full of such dreadful speculations. The late Michael Crichton dreamed up the world of Jurassic Park, in which scientists possessing knowledge and skills to bring dinosaurs back to life were not prescient enough to keep the wild creatures from reproducing out of control. In a later novel, *Prey*, Crichton imagined a swarm of self-replicating nanobots that kill and colonize human bodies until countertechnological means are found to destroy them.

Fiction pushes the envelope of technological futures to unrealistic, sometimes absurd extremes. Yet, as we saw in chapter 2, prudent societies have long invested in protecting themselves against future harm. Insurance and risk assessment evolved to make sure that worst-case scenarios would not descend without warning and leave people destitute—including risks associated with technological progress. To be sure, it is relatively easy to calculate and guard against the statistically predictable prospect that farmers will be injured by drought, merchants by storm, and homeowners by earthquake, fire, and flood. It is harder to quantify the catastrophic possibilities associated with many new and emerging technologies: for example, that synthetic pathogens will escape, overwhelm human immune systems, and cause millions of deaths; or that nanobots will replicate unchecked and kill their makers; or even that climate change will devastate world agriculture on an unprecedented scale. Yet efforts have been made to envision and ward off even those relatively far-fetched possibilities, though almost by definition they cannot be foreseen by eyes and minds situated in the present.

In this chapter, I review the major procedural mechanisms by which modern societies have tried to reclaim the technological future as a governable space, one that can be safely inhabited because it conforms to society's ideas of the good and does not evade the possibility of control. Each mechanism, however, raises its own dilemmas of power and deliberation. Foremost is the question of inclusion. Who gets to participate in imagining the futures toward which eager technology producers wish to steer the world? Coupled to that most basic question of democracy are more specific puzzles about the design of institutions that might enable a genuinely collective reflection on technology's potential. Several strategies have attracted policymakers' attention in recent decades, and each deserves closer inspection: technology assessment, or the systematic mapping of alternative technological pathways, and its offshoot constructive technology assessment; ethical analysis; and methods of public engagement that aim to reinvigorate deliberative democracy.

TECHNOLOGY ASSESSMENT

Looking back, 1972 was a year of world-changing political events such as the winding down of the Vietnam War and the birth of Bangladesh. Importantly for humankind, it was also a milestone year for technology: the United Nations Conference on the Human Environment in Stockholm, Sweden; the first successful lab recombination of DNA; and the launch of *Apollo 17*, the last manned flight to the moon. Hardly noticed among these great events was a law enacted by the U.S. Congress in October of that year. The Office of Technology Assessment Act[3] was the brainchild of influential academic science advisers, led by Harvey Brooks, physicist and engineering dean at Harvard.

These men were determined to equip federal lawmakers with the capacity to make rational policy in the face of rapid technological change. The act's declaration of purpose stated, "[I]t is essential that, to the fullest extent possible, the consequences of technological applications be anticipated, understood, and considered in determination of public policy on existing and emerging national problems."

To serve Congress's need for policy-relevant advice, the law created a new administrative body, the Office of Technology Assessment (OTA). Legislators worried, though, that in bowing to the need for better information they might be delegating too much power to experts or to runaway political factions that might co-opt the OTA's expertise for their self-serving purposes.[4] Compromise after compromise trimmed the sails of the new agency before it was launched. Congress ultimately settled on an institutional design that was small, politically balanced, organizationally flat, and heavily dependent on outside consultants. A twelve-member Technology Assessment Board (TAB) governed the OTA, composed of six members each from the Senate and the House of Representatives, and evenly representing the two major parties. The board appointed the OTA director and approved every study the agency conducted. In-house OTA staff numbered fewer than two hundred permanent employees, about three-quarters of whom were researchers; otherwise, studies drew on a network of temporary consultants with relevant expertise from industry and universities. How did an organization whose sole purpose was to bring reason into politics function inside the rough and tumble of the legislative process? The OTA's twenty-two years provide some insights into that precarious balancing act and, more generally, into the limits of injecting a technically informed voice into American political decisionmaking.

Tensions surfaced fast. The final years of the Nixon administration were a time of sharp political polarization that divided liberals and conservatives on many issues with a technological component, such as arms control, pharmaceutical drug safety, and environmental protection. An air of partisanship, undermining the promise of balance, hung over the OTA's operations from the appointment of its very first director, Emilio Q. Daddario, a former Democratic congressman from Connecticut who had been the chief advocate of the law creating the agency. Daddario was seen as too deferential toward the largely Democratic politicians who controlled the TAB under its powerful first chair, Senator Edward M. Kennedy of Massachusetts. By the time Daddario stepped down as director, the agency was caught in a whirlpool of rancor; Republican TAB members charged that Kennedy was attempting to take over the agency and reconstitute it as an extension of his own staff and political agenda.[5]

Subsequent OTA heads steered the agency away from partisanship toward political neutrality and wider buy-in from congressional committees. John H. Gibbons, the third and longest-serving director, significantly reshaped the OTA's identity in his twelve years at the helm. Abandoning the largely single-committee sponsorship of its work in the 1970s, the OTA had moved by the 1990s to averaging almost three sponsors per study. A crucially important strategy for maintaining neutrality was to avoid making direct policy recommendations. Instead, OTA offered alternatives, laying out the pros and cons of each option. Asked to consider improving automobile fuel economy in 1991, for example, the OTA considered not only raising standards but other options such as economic incentives and technological design that would lead to more efficient fuel use. Gibbons scored an early political success by heading off a move to eliminate the OTA in the deregulatory fervor of the 1980 election that swept Ronald Reagan into the White House. The next fourteen years brought relative calm, but the agency did not survive a second major upheaval, the so-called Republican revolution of 1994.

Led by Speaker of the House Newt Gingrich, the 1994 midterm elections secured Republican control of both houses of Congress for the first time in forty years. Gingrich's inspired campaign pledge, "Contract with America," featured a ten-point plan for cutting taxes, reducing the federal budget, and eliminating a raft of social welfare programs installed by Democrats in preceding decades. Flushed with victory, the new majority in Congress saw a symbolic need to cut its own spending, and when the OTA again appeared on the chopping block there was little enthusiasm and less will to save it. The OTA became in a sense the victim of its own success, neither big enough to matter nor political enough to generate an impassioned defense. According to one analyst,

> In the end, OTA's funding was eliminated not because it was too large, but because it was so tiny. Terminating the agency did not require Congress to absorb the loss of a large or well institutionalized operation. It offered a small real contribution to budget reduction and seemed to provide a larger symbol of congressional commitment to reducing the size of government.[6]

The OTA disappeared as a live presence on the American policy scene, but what ended was its funding, not its authorization. As a result, the agency survives in the imagination of senior American leaders of science and engineering as a ghostly reminder of golden days, a presence that could be summoned forth again by a legislature with the courage to stand up for reason in tackling urgent issues such as climate change. Thus, *Restoring the Foundations*, a 2014 report by a prestigious committee of the American Academy of Arts and Sciences, chaired by Neal Lane, former National Science Foundation director and prominent science policy adviser, notes, "The lack of a mechanism in Congress to address national science and technology issues and to coordinate policy with the President remains a major policy issue." The text goes on to state that the OTA's "authoritative analysis was critical to legislative decision-making related to science and technology."[7]

With twenty years of hindsight, however, it remains unclear how far the OTA fulfilled its mission of providing "authoritative analysis" to Congress, and still less clear how far its policy reports influenced legislation. Three cases can be taken as illustrative of the OTA's contributions to technology policy, as well as its overarching aims of democratizing technical decisionmaking and placing the nation's technological future on a sound footing. The cases concern arms control, biotechnology, and the practice of technology assessment.

The Star Wars Debate

The first example comes from defense policy, in particular, the OTA's assessment of the Strategic Defense Initiative (or "Star Wars") proposed by President Reagan in 1983. A year later, the OTA, working together with the antinuclear Union of Concerned Scientists, published a background paper showing that the shield advocated by the president and his advisers was technically unworkable—a "pipe dream," as the prominent *New York Times* columnist Tom Wicker dismissively dubbed it.[8] The paper drew heavily on the work of the MIT physicist Ashton Carter, a former OTA employee and leading nuclear analyst, whose involvement with that project propelled him to a professorship at Harvard's John F. Kennedy School of Government and eventually to the top position in President Barack Obama's Department of Defense. But while the liberal establishment applauded the OTA's verdict as a superb example of science trumping politics, the paper elicited howls of rage from Reagan supporters, including the Heritage Foundation and the *Wall Street Journal*, who saw it as bad science driven by politics and an irresponsible disregard for national security.[9] Opponents charged that the OTA committee that had commissioned the report was stacked with Kennedy and Johnson administration ideologues, such as McGeorge Bundy; as the paper's primary architect, Carter drew special ire for having made public data he had obtained under a special security clearance.

Supporters of the OTA analysis countered not only that the proposed missile shield was unworkable but that it would ignite Soviet paranoia and destabilize the hard-won "balance of terror" that kept the vast nuclear arsenals of the two global adversaries under control. Later, after the fall of the Berlin Wall and the dissolution of the Soviet Union, some conservative analysts argued that the dynamics of fear had actually worked to favor the United States—engaging in a competitive, defensive arms race had exhausted the Soviet treasury and led to the downfall of communism. These debates are likely to remain unresolved for decades, until enough time passes and enough classified documents become available for meticulous historical research. It is fair to conclude, however, that the OTA's intervention in this case failed to carve out the space of neutral expertise that its designers had hoped for. The background paper on Star Wars became one more loud, discordant note in the ongoing cacophonous debate on Cold War defense policy. Trusted and embraced by liberals, but dismissed as a political ploy by conservatives, it would be difficult to argue that the OTA analysis significantly advanced the cause of informed, deliberative democracy.

The OTA and Biotechnology Policy

The archived website of the OTA lists twenty-eight reports under the heading "biological research and technology," stretching from a study of the "impacts of applied genetics" in 1981 to a posthumous report on "federal technology transfer and the human genome project," released in September 1995 by the already shuttered agency. In between, the OTA produced reports on a wide array of policy-relevant topics, such as the workplace impacts of genetic screening, forensic DNA testing, and the position of the U.S. biotechnology industry in relation to developments in other countries. Five reports published between 1987 and 1989 covered "new developments" in biotechnology, from patenting to public perceptions. But what impact did this impressive record of approximately a report a year have on the ethics, politics, and policy of this emerging technological sector, arguably one of the most profitable as well as controversial in recent U.S. history?

One of the most significant regulatory decisions to emerge from the Reagan years was to go easy on biotechnology. Just as the U.S. Supreme Court held in 1980 that patenting life forms was no different from patenting anything else "under the sun made by man," so the president's Office of Science and Technology Policy decided in June 1986 that there was nothing new enough about genetic engineering to warrant legislation specifically targeting biotechnology. The Coordinated Framework for Regulation of Biotechnology concluded that, by and large, existing laws provided adequate jurisdiction to cover the products of biotechnology.[10] Federal policy could streamline operations through such measures as designating a lead agency for each class of products, standardizing definitions of key terms such as "new (intergeneric) organism" and "pathogen," and providing for timely exchanges of information between agencies.

This relatively laissez-faire approach to biotechnology positioned the United States as an early industry leader in both agricultural (green) and pharmaceutical (red) biotechnology. As we saw in chapter 5, however, as soon as U.S. plant biotechnology sought to export its products across national borders, it encountered wave upon wave of resistance, especially in Europe, where regulation was founded on a more precautionary approach and wider commitment to organic agriculture. Even American consumers proved to be less than totally thrilled with Monsanto's brand of market imperialism. Consumer refusal to tolerate GM foods under the organic designation, as well as continuing state-by-state campaigns for clear labeling of food containing GM ingredients, testified to an environment of suspicion and, for many, outright rejection.

There is little to suggest that the OTA foresaw this trouble in the making, or that it effectively advised Congress on potential alternatives to the Coordinated Framework. In its 1988 report on field-testing genetically modified (GM) organisms, a study in its "new developments" series, the OTA examined many of the technical questions that later became publicly controversial. For example, the report noted the potential for gene flow across biologically similar organisms, and yet it somehow represented them as harmless by stressing the extremely low probability of catastrophic accidents. Unknown unknowns (see chapter 4) were not in the OTA's sights, despite the novelty of the technology and the complexity of the living systems opened up to reengineering. With regard to the Ice Minus bacterium, which caused such controversy in California and which in the end proved to be commercially unviable, the OTA concluded there was little cause for concern: "Several different studies suggest, however, that even under a long chain of worst-case assumptions (many of which contradict known facts) the alteration of climatic patterns through large-scale agricultural applications of ice minus bacteria is not likely. Many of these assumptions, however, could benefit from being tested by further research."[11]

Summing up, the OTA charted three options for Congress in reviewing applications for planned introductions of GMOs into the environment. The first was business as usual—that is, to require nothing different from any other premarket review, even though the report acknowledged that there might be "problems stemming from the ability of living organisms to grow, reproduce, or transmit genetic material to nontarget species."[12] The second, most detailed option, apparently the one the OTA expert group liked best, was to subject all applications to prior assessment, but to assign each to a risk class that would determine the scope and intensity of review. The third and least-favored option was to subject each application to maximum scrutiny.

If one of the OTA's functions was to turn the balance wheel of expertise toward the more democratic forum of Congress on value-laden issues, its biotechnology reports do not seem, in retrospect, to have fulfilled that mission. The timing and content of the OTA's policy proposals fell strikingly short of foreseeing future difficulties, let alone mobilizing congressional opinion to think independently about biotechnology. By 1988, the White House had already issued the Coordinated Framework, strongly endorsing what was in effect the OTA's first, least intrusive option. This proposal became and remains U.S. policy. Yet the specificities of biological organisms that the OTA noted did not fade away. For example, the effects of huge swaths of GM corn on nontarget species such as the monarch butterfly continue to resurface in ecological science despite industry efforts to dismiss them as ungrounded.[13] Similarly, interspecies transmission of genes periodically returns to haunt biotechnology, as when a controversial study by Ignacio Chapela of the University of California at Berkeley purported to show that genetically modified traits from U.S. corn had contaminated locally valued native species in Mexico.[14]

Constructive Technology Assessment

The OTA's influence outside the United States may have outstripped the effects it had on American domestic policy, although even here the agency served more as inspiration for a way of thinking than as a model of best practice. During the 1980s, numerous European nations adopted the principle of technology assessment along with the idea of a specialist agency to assist parliamentarians on matters involving a scientific or technological component. Denmark and the Netherlands were among the first to set up offices similar to the OTA, in the mid-1980s in both cases. Britain established the Parliamentary Office of Science and Technology (POST) in 1989 to offer information and briefings to committees and individual members of Parliament. The inclusion of science in POST's mandate speaks to the high value placed on research and research policy in Britain, and, unlike other comparable agencies, POST defines its role more in terms of providing information than in terms of technology forecasting or policy advice. A Europe-wide body, the European

Parliamentary Technology Assessment network, offers opportunities for information sharing and joint projects among the varied national organizations.

Proliferating across diverse bodies and political cultures, technology assessment developed along divergent lines, following nationally specific traditions of connecting technical analysis with democratic deliberation. One such strategy, first developed in a 1984 policy paper by the Netherlands Office of Technology Assessment (NOTA), came to be known as constructive technology assessment (CTA). NOTA's point of departure was the observation that citizens' views should be built into the design of technologies. This can be done only through participatory processes, such as consensus conferences, in which prototypical citizens actively articulate for policymakers how they wish a given technology to progress. CTA imagines a need for constant communication between technology and society so that the "historical experience of actors, their views of the future, and their perceptions of the promise or threat of impacts" are continually fed back into the progress of technology.[15] This vision struck a responsive chord in a country with a strong tradition of consensus-based policymaking, and a variety of Dutch agencies and trade associations adopted procedures for citizen engagement on issues such as health care policy.

In its design, CTA looks fundamentally different from the OTA's approach, which relied on a range of experts to represent diverse viewpoints but did not include a space for direct inputs from citizens or other actors. One may question, however, whether even CTA opens up a truly egalitarian space of the imagination, allowing varied players to participate in the design phase of technological trajectories, or whether it is more a pragmatic exercise in securing public buy-in for choices made in advance by government and industry. As an offshoot of technology assessment, CTA may well concentrate technology assessment's shortcomings as well as its virtues as a mode of collectively imagining and adapting to the future.

The heart of the problem lies in the fact that CTA sits squarely on the side of policy, not politics. The term "constructive" signals a positive commitment to technologies, and very occasionally to pull back from commitments, that are not themselves open to questioning. CTA is evolutionary, not revolutionary. Its procedural approach affirms its sensible and progressive leanings, but it operates more as a technique of implementation than as a forum for wrestling with the ethics of grand schemes for technological improvement. In its very commitment to dialogue, inclusion, and informed debate, therefore, CTA seems ill suited for wide-ranging, acrimonious disagreement about how humans should deploy their ever-expanding technological powers to achieve ends that exceed the reach of politics as usual—such as the decarbonization of the oil-based economy, enhancement and interconnection of human brains, bringing extinct species back to life, or reengineering the skies to cool the planet. Two green energy initiatives, both failures, offer closer looks at how the politics of technology policy sidelines public conversations, essentially eliminating the possibility of constructive public engagement. Both illustrate a dynamic of close collaboration between modern states and their major industries, to the exclusion of disaggregated, small-scale political actors.

The first is the story of the zero-emissions vehicle (ZEV), or electric car, in the United States. In 1990, the California Air Resources Board demanded that major car manufacturers, including General Motors, convert 2 percent of their fleet to ZEVs over the next eight years in return for continuing to sell gasoline-powered cars. In theory, CTA actors could then have gone about creating the multilevel dialogue needed to coordinate how such a mandate might be turned into material reality. In practice, however, as suggested in the 2006 film *Who Killed the Electric Car?*, General Motors undermined consumer demand and quite possibly colluded with federal and state regulators to induce California to scale back its ZEV mandate by the turn of the century. In the documentary, crushed electric vehicles symbolize the insuperable power of business-government relations in comparison with that of potentially interested publics.

The second story concerns Desertec, an ambitious plan to cover the Sahara Desert with solar panels to generate electricity for North Africa and Europe at an estimated cost of some $550 billion. Formulated in the early twenty-first century, the plan initially appeared technologically feasible and generated much excitement among big German industrialists such as Siemens and Bosch, but it failed to win the expected financial backing and was eventually abandoned when Siemens and other key players lost interest. After-the-fact verdicts noted that promoters had paid little attention in their early thinking to the local politics of generating electricity in North Africa for export to Europe. The region's growing turbulence, fed in part by rising Islamic fundamentalism, evidently caught the energy technocrats by surprise. Eventually dismissed as "utopian" and "one-dimensional," Desertec illustrates the limits of top-down conceptual planning and, more crucially, the absence of entry points for the involvement of concerned publics, exercising social intelligence, at the earliest stages of technology policy.

INVENTION'S ETHICAL QUANDARIES

Transformative power of a kind that humans cannot exert, and yet would dearly love to, once was attributed to God alone.

Only divine knowledge, our predecessors thought, could eliminate the ancient evils of disease and injustice, probe into the
recesses of the human mind, and guide the functioning of whatever moves upon the Earth or around it. Science and technology
have brought many of these superhuman capabilities within our
reach, in effect enabling mere mortals to "play God." Yet anxieties run deep about hubris and ignorant overreaching. They surface quickly when technological developments seem to unleash
the potential for catastrophic harm, threaten to turn nature into
artifice, or, as discussed in chapter 5, destroy essential features
of humanness that people believe should be shielded against
even well-intentioned refashioning.

One response from states has been to delineate, through policy, areas that should be kept clear of technological intrusion.
Throughout the twentieth century, countries came together
around agreements not to pursue developments that most found
abhorrent. International treaties set limits on technologies,
and technologically supported social behaviors, that threaten
human survival and the planetary environment. Accords
included nuclear disarmament and nonproliferation; bans on
chemical weapons, land mines, and persistent organic pollutants; the phaseout of ozone-depleting chemicals; the protection
of endangered species; and the preservation of biodiversity and
singular natural sites against indiscriminate industrialization.
Most of these actions focused on physical damage to health,
safety, or the environment, and, at the limit, to human existence. More rarely, agreements emerged not to undertake technological activities that might inflict irreversible moral harm.
The de facto global bans on cloning human beings and modifying heritable genetic characteristics rest on just such widely
(though not universally) shared intuitions concerning frontiers
that biotechnology should not cross.

New and emerging technologies raise many ethical concerns that are profoundly troubling but do not rise to quite the level of existential threats posed by nuclear weapons or human cloning. How can societies make sure that, in their headlong rush toward new knowledge and novel capabilities, science and technology do not lose touch with public values and trample on real, though inarticulate, moral sensibilities? The growth of public ethics bodies over the past half century offers a partial answer. Many nations have adopted ethical deliberation as a policy practice that aims to synchronize technological progress with societal values. Like the varied forms of technology assessment, these processes, too, offer distinctive virtues and limits in securing democratic control over technological futures.

Ethical deliberation on technology began in many Western societies around advances in biomedicine and biotechnology, a move that gained urgency through the disclosure of horrific abuses by German medical science during the Second World War. The Nuremberg Doctors' Trial, carried out by military courts after the principal war crimes trials, brought twenty-three prominent physicians and officials to the dock for unlawful medical experimentation and mass murder. One result was the adoption of the Nuremberg Code, a set of principles governing research on human subjects. The first is the principle of informed consent, which serves as the cornerstone for contemporary biomedical ethics. It states that, in all experimental contexts, "the voluntary consent of the human subject is absolutely essential."[16] Major research institutions throughout the world today have standing committees—known in the United States as institutional review boards (IRBs)—to ensure that researchers obtain valid informed consent before embarking on potentially dangerous studies.

The principle of informed consent and its associated safeguards apply most immediately between individual researchers and the subjects of their experiments. That narrow reading of the Nuremberg Code, however, sits within a broader consensus forged during the postwar period that societies have a right to hold technologies and their sponsors ethically accountable, especially when technological advances threaten major disruptions in intimate social relations or the distribution of risks and benefits. There is no single code or constitution that formally announces this principle, but many discrete institutional developments over the past half century have endorsed ethical forecasting and care as elements of what we might regard as a kind of de facto supranational constitutionalism. The core commitment is to the prior assessment of ethical implications and consequences, complementing conventional technology assessment's preoccupation with impacts on health, safety, economy, and environment.

Institutions that enable societies to reflect on the ethical dimensions of technological innovation take two major forms: highly visible, politically appointed bodies, often speaking directly to matters of national policy; and barely visible, managerial bodies looking after the day-to-day dynamics of research. National-level committees may enjoy some agenda-setting powers, though they can also be called upon by politicians to address urgent new questions. Such committees, moreover, occupy very different positions vis-à-vis the decisionmaking arms of their respective national governments, and their powers are correspondingly shaped and circumscribed.

In the United States, national ethics bodies began operating in 1974, when Congress asked the Department of Health, Education, and Welfare to establish the National Commission for the Protection of Human Subjects of Biomedical and Behavioral Research. That commission's most significant contributions included the criteria for brain death and the establishment of institutional review boards to approve research on human subjects. Congress mandated the creation of several bioethics bodies well into the 1980s, but legislative leadership ended with the failure of the short-lived Biomedical Ethical Advisory Committee, operational between 1988 and 1990, whose evenly balanced membership, drawn equally from the two legislative chambers, proved unable to work around the bitter stalemate of U.S. abortion politics. Since 1996, starting with the Clinton administration, national bioethics bodies have operated under explicit presidential directives.

Whether serving at the president's pleasure, or that of Congress, U.S. national ethics commissions are patently subject to political pressures. Committees formed by the White House in particular are led by people who tend to be broadly in sympathy with the president's political and policy preferences. Thus, Dr. Leon Kass, who chaired the Bush era President's Council on Bioethics from 2001 to 2005, was known for his conservative leanings on biomedical research. He famously articulated the "yuk reaction," formally known as the "wisdom of repugnance," as a test for directions that society should hesitate to pursue.[17] By contrast, President Obama turned to Professor Amy Gutmann, a political philosopher of impeccable liberal credentials, to lead his Presidential Commission for the Study of Bioethical Issues. Under Gutmann's direction, the commission deferred to science to define the promise of new technologies, while asking for caution largely in areas where risks seemed self-evident. For example, when the noted scientist-entrepreneur J. Craig Venter claimed that his company had created a new synthetic microorganism, the commission not only concluded there was no need for regulation, but recommended that a public watchdog be appointed to make sure the press did not purvey misleading fears—including the use of inflammatory language such as "playing God."[18]

A far less prominent role is played by the cluster of ethics committees that supervise the conduct of research at leading U.S. universities. These include IRBs responsible for human subjects research, Institutional Animal Care and Use Committees (IACUCs) responsible for animal experimentation, and most recently Embryonic Stem Cell Research Oversight (ESCRO, or sometimes just SCRO) bodies created in response to American public concern over the derivation of stem cells from human embryos. Unlike the national ethics committees—whose judgment and reassurance are often in demand after seemingly game-changing technological breakthroughs such as the birth of Louise Brown, the first test-tube baby, or of Dolly, the cloned sheep—IRBs, IACUCs, and ESCROs serve as almost invisible handmaidens to the research enterprise. They are entrusted with preventing flagrant ethical violations, but at the same time are expected not to block the progress of science as scientists see it. They are appointed by university administrators drawing on in-house faculty and staff, but (unlike those of national committees) their members are chosen for their expertise and are not subject to obvious political screening or requirements for ideological balance.

These committees, too, operate under constraints that limit the scope of their ethical imaginations. Their position is that of gatekeeper, with the power to block lines of research that seem too risky for reasons the committee is responsible for analyzing. Ethics committees are intensely aware of possible reputational damage to their home institutions, not to mention the drastic prospect of fines or suspension of colleagues if public funding bodies uncover serious transgressions. Though such considerations ensure extreme care in the application of guidelines, oversight frequently reduces to a fairly mechanical process of ensuring that all the right boxes have been ticked. On the flip side, ethics committees at leading research universities are also sensitive to the pressures on their preeminent scientists. High-caliber science these days is fiercely competitive. Prestigious publications, grants, and prizes hang not only on getting big results but on getting them first. By composition and mandate, ethics committees tend to be sympathetic to their scientist colleagues' ambitions and priorities and reluctant to raise barriers against what they see as society's legitimate interest in rapid scientific progress. In turn, this means that ethics committees are not best situated to question the fundamental purposes of new lines of scientific research or innovation; for the most part, they accept those purposes as given and see their role as being science's helpers rather than supervisors or adversaries.

Although U.S. technology development programs have repeatedly set aside funds for public ethical deliberation, no single model has emerged for how best to carry out such reflection. Perhaps best known is the Ethical, Legal, and Social Implications (ELSI) Program of the Human Genome Project, launched in 1990. ELSI initially operated as a centralized funding program administered by the National Institutes of Health and the Department of Energy. Its mission was to "anticipate problems and identify possible solutions," mostly by funding investigator-initiated proposals. Although ELSI research continues at the National Human Genome Research Institute, project funding was decentralized in the late 1990s and placed under the supervision of the ELSI Research Advisors Group.[19]

ELSI's history suggests that neither the scientific community nor its federal sponsors were satisfied with the NIH's initial approach to funding ethics research, where proposals and priorities came largely from investigators pursuing their own interests. Subsequent initiatives, such as programs for funding nanotechnology and synthetic biology, opted for more tightly controlled approaches to research on the ethical and social dimensions of their respective domains. Not surprisingly, the recommendations coming from such initiatives have highlighted actions scientists can take on their own, under such rubrics as "responsible innovation" and "anticipatory governance." The presumption is that scientists immersed in the research process are best placed to understand and resolve any dilemmas associated with their work, possibly with the aid of an in-house ethical adviser or two. Ethical analysis is conceived, then, as serving instrumental rather than broadly democratic ends, including the government's need to reassure concerned publics that moral risks are under control or to develop policy on specifically troublesome issues such as genetic privacy.[20]

Ethical analysis that does not advance such instrumental purposes tends to be downgraded as not worthy of public support. In 2010, for example, an unusually public fight erupted between the National Science Foundation and Paul Rabinow, a noted anthropology professor at the University of California at Berkeley, who had been hired to study the ethical dimensions of the Synthetic Biology Engineering Research Center (SynBERC), headquartered at the university. An NSF review concluded that some aspects of the research conducted by Rabinow and his colleagues "appear to be primarily observational in nature rather than proactive and developmental."[21] Decoded, this language meant that NSF was dissatisfied with SynBERC's progress toward developing security guidelines for potentially dangerous biological research, and that goal took precedence over exploring other ethical dimensions of synthetic biology.

Rabinow effectively lost control of Thrust 4, the portion of the project devoted to studying ethical issues, and leadership for the effort passed instead to an insider scientist, Drew Endy. Rabinow stayed on for some months as a researcher but eventually resigned from SynBERC, blaming what he said was the scientists' indifference to their "responsibility to larger society, which is funding them, by entrusting them to manipulate life."[22]

The design and operation of ethics bodies differ widely across as well as within nations, creating different mixes of blockages and opportunities for the inclusion of public values. On the whole, there is greater input from lay publics in both German and British approaches toward dealing with morally troubling technological developments, particularly on matters involving the life sciences and technologies. In Germany, parliament rather than the executive controls the national institution of ethical deliberation. The Ethics Council Act of 2007 established the German Ethics Council (*Deutscher Ethikrat*), whose duties are to "pursue the questions of ethics, society, science, medicine and law that arise and the probable consequences for the individual and society that result in connection with research and development."[23] Through a mix of opinions, recommendations, and annual reports, the council covers the moving frontiers of biomedical research, probing especially those issues that seem to expose gaps in the coverage of existing laws. For example, the Council's 2011 report on human-animal chimeras recommended that the Embryo Protection Act be amended to forbid the implantation of a "cybrid" (consisting of a human nucleus transferred into an animal egg) in a human uterus.[24] That possibility was neither foreseen nor guarded against when Germany enacted the embryo law in 1990.

In contrast to Germany, which opted for a formal and centralized coupling of science, ethical analysis, and lawmaking,

Britain went to the other extreme in adopting a loosely structured and unofficial approach to public ethical deliberation. The Nuffield Council on Bioethics is the most prominent body offering ongoing deliberation and advice on ethical issues involving bioscience and biotechnology in Britain. It was established in 1991 by its parent Nuffield Foundation and is supported by grants from the Wellcome Trust, a major UK source of biomedical research funding, and also by the government's Medical Research Council. Anticipation features prominently in the council's mission statement just as it did for ELSI in the United States. The council's first objective is "to identify and define ethical questions raised by recent advances in biological and medical research in order to respond to, and to anticipate, public concern."[25] Here the assumption is that an expert body can see farther ahead than the lay public, and its advice in turn can help policymakers allay potentially crippling public concerns.

Though not convened by law or executive mandate, the Nuffield Council subscribes to common democratic norms to safeguard its legitimacy. Members serving up to two three-year terms are selected through public advertising of openings and through professional networks. Like any comparable public body, the council seeks to maintain a reasonable balance across gender, ethnicity, and areas of expertise. It selects its study topics through consultation, but it also invites suggestions for future work from the public on its website. All these procedures mimic those any official state agency might adopt, and yet no political institution or actor is responsible for the council's ultimate reports.

In sum, as these cases show, the principle of ethical deliberation on technological futures has become commonplace in industrial societies. There is a transnational commitment to anticipating and holding at bay possible moral harms through timely forecasting and precautionary action. Yet the most firmly institutionalized modes of ethical analysis seem all too often to fall together with dominant ideologies, as in the pronouncements of politically appointed national ethics committees or the domesticated and largely rule-following work of many lower-level institutional ethics bodies. The professionalization of "ethics" in committees charged with supervising research conduct thus raises troubling questions about who controls technology.[26] The proliferation of such bodies seems to have supplanted democratic imaginations of good technological futures—what we might call public sociotechnical imaginaries[27]—with a narrower technocratic vision that puts the brakes on research only when experts agree there are manifest risks to institutional reputation or to subjects' health and safety. Can efforts to engage lay publics directly in debating their technological futures do any better?

PUBLIC ENGAGEMENT: PANACEA OR PLACEBO?

States have long recognized that technology is a powerful resource for securing the assent of the governed, the foundation on which democracy ultimately rests. Technological demonstrations, especially when a nation is under attack, can persuade citizens that their government is capable of safeguarding the people against death and devastation. The flip side is that a big technological failure can strain or undermine trust in the state. Examples include the 1984 Bhopal disaster in India, the 1986 Chernobyl nuclear plant accident in the Soviet Union, and the second "mad cow" crisis of the 1990s in Britain, in which a deadly degenerative brain disease spread from cattle to humans despite confident expert assurances that such a transfer between species was extremely unlikely to happen. In all these cases, states were left to pick up the shards of shattered public confidence, with varying degrees of success. These cautionary tales have entered the discourse of democracy, underscoring a need for prior consent when governments embark on technological projects that could cause great harm. Public participation or, in more recent terminology, public engagement in technological decision-making offers an opportunity for citizens to work together with scientists, engineers, and public officials to envision more inclusive technological futures. What do such exercises look like in practice, and to what extent have they permitted citizens to inject their values into the courses of technological development?[28]

Attempts to involve the U.S. public in decisions about the control of technology traditionally played out under the umbrella of law, using the time-tested mechanisms of the adversary process. As long ago as 1946, Congress passed the Administrative Procedure Act (APA), a milestone in modern lawmaking that enshrined the principle of public consultation as an element of any exercise of federal regulatory power. During Senate debate before the APA was passed, Senator Patrick McCarran of Nevada, one of the law's chief sponsors, observed that a "fourth dimension" of governance—the administrative—had opened up, a dimension the Constitution did not foresee or explicitly allow for. Tellingly, McCarran referred to the APA in constitutional language, as "a bill of rights for the hundreds of thousands of Americans whose affairs are controlled or regulated in any way by agencies of the Federal Government."[29]

If the APA relaid the constitutional groundwork for a democracy increasingly run by expert federal agencies, then the burst of health, safety, and environmental legislation of the 1970s built upon that foundation a superstructure that dominated the governmental skyline of the later twentieth century. Though conservative critics of big government typically dismiss it as government run amok, that legislative flowering expresses a too often overlooked confidence in citizens' capacity to understand and influence even the most arcane aspects of federal policy. Those laws conceived of citizens as being not necessarily *knowing*, but knowledge-*able*—that is, capable at need of acquiring the knowledge needed for effective self-governance. This idea of an epistemically competent citizen runs through American political thought from Thomas Jefferson to John Dewey and beyond. The fervor for deregulation and neoliberalism that gathered steam in the 1980s, then, challenged not only government's ability to solve public problems but also the public's ability to hold technology accountable to common values. In effect, the deregulatory turn ceded to the private sector, and its most powerful representatives, the right to set agendas, develop expertise, and make design choices without the continuous critical monitoring from below enabled by decades of democratizing legislation.[30]

A hundred years of U.S. administrative practice navigated between a vision of technological democracy in which states regulate innovators under legally enforceable safeguards, and a supposedly more liberal vision in which technology producers envision what publics want, and deliver the goods, subject mainly to the laws of the market. It is tempting to see this period as one of declining democracy and increasing privatization and fragmentation of collective reasoning.[31] Libertarians, however, see the new possibilities for self-expression and self-governance created by the digital media as a welcome opener for direct democracy and an improvement over state paternalism and capture of deliberative spaces by powerful interest groups, such as agribusiness, big pharma, and even orthodox medicine. Such emancipatory enthusiasm frequently accompanies the advent of new technologies; as discussed in chapter 6, however, it remains to be seen whether the democratic promises of the Internet age will lead to more or less control of individual and collective destinies.

Participatory developments in the United States contrast markedly with those in nations where the state historically enjoyed greater authority as custodian and enactor of what Jean-Jacques Rousseau termed the "general will." The expansion in participatory rights for citizens that characterized the American midcentury either did not occur in much of Western Europe or occurred in less far-reaching form than in the United States. Authoritarian states and developing countries were still slower to adopt procedural reforms that give voice to citizens affected by technological choices largely controlled by governments, for example, in agriculture, energy, economic, medical, or industrial policy. The worldwide protests against the introduction of agricultural biotechnology (see chapter 4) can be seen in this light as a global case study in direct democracy: a series of demonstrations through which small farmers, environmentalists, development critics, and other skeptics around the world began challenging not only a particular form of technological innovation but the state-centric politics undergirding its global spread.

Turn-of-the-century developments in Britain offer especially instructive contrasts with the U.S. case, illuminating the pros and cons of divergent approaches to public participation. UK administrative practice historically relied on informal consultation among tightly networked elites as a basis for policy. Until very recently, formal mandates to open up decisionmaking were few and far between. Whereas the U.S. Freedom of Information Act dates back to 1966, with significant expansion in 1974 in response to the Watergate hearings, a law granting more limited access to government information was enacted in Britain only in 2000. Consultation remains fundamental to policymaking, but for the most part the state decides who should be consulted, as if fearing the turmoil of letting an unscreened multitude in. Accordingly, some UK analysts have called for more "uninvited participation" aimed at drawing a wider range of public views and values into policymaking.[32]

One ambitious initiative to engage more citizens in a public debate on technology reveals the tensions created by attempts to open up a traditionally closed process. This was GM Nation?—a yearlong effort by the UK government to elicit public opinion on the introduction of GM crops. Carried out on the advice of the short-lived but influential Agricultural and Environment Biotechnology Commission (AEBC) in 2003,[33] GM Nation? involved over six hundred public meetings and events around the country, allegedly to inform a major new policy statement on GM crops. The verdict appeared clear: only 2 percent of the surveyed public favored GM crops, and 95 percent worried about contamination of non-GM agriculture. Most participants expressed considerable uncertainty about the benefits of agricultural biotechnology and commented on the absence of independent oversight of a domain long seen as captive to major industry players. Yet although GM Nation? produced unambiguous results, was closely watched by government regulators, and spawned much press commentary,[34] it was not embraced as a model for future deliberation. Biotechnology advocates dismissed it as a forum for self-selected antitechnology forces, while opponents accused the government of having made up its mind to follow a pro-GM policy regardless of the dialogue's outcome. A government report on "lessons learned" from the debate was decidedly equivocal, highlighting that it had failed to integrate public opinion with concurrent scientific and economic technology assessments.[35] In other words, GM Nation?

was deemed an official failure because it did not bring the public around to agreeing with the state's preferred policy position.

In longer retrospect, Britain's unique experiment with drawing the national public into a dialogue on the ethical and social implications of a new technology comes across as too little, too late, and too idiosyncratic. It attempted to retrofit a historically closed and inward-looking policy culture, reliant on negotiation among trusted insiders, with an uncommonly open and inclusive exercise in self-governance. That effort, moreover, came after the issue was already thoroughly polarized and the towering genie of dissent had escaped from the puny bottle of policy containment. Under those circumstances, GM Nation? was almost doomed to fail. A group of authors from Lancaster University, including Robin Grove-White, an AEBC member, concluded that the "agricultural GM experience evidenced a tendency that when faced with new situations and technologies, regulators turn to assessment frameworks developed for previous technologies and tied into existing debates." That recourse to fighting the last war prevents, in the authors' view, a properly "searching, socially realistic analysis of the distinctive character and properties of particular new technologies."[36]

CONCLUSION

Technologies, we have seen, are not merely tools for achieving practical ends but devices with which modern societies explore and create potentially more liberating and meaningful designs for future living. Through technology, human societies articulate their hopes, dreams, and desires while also making material instruments for accomplishing them. Collective visions and aspirations, moreover, change and evolve as societies become habituated to new technologies and use them to pursue altered understandings and purposes. Technological choices are, as well, intrinsically political: they order society, distribute benefits and burdens, and channel power. It is surprising, then, that technology and democracy largely steered clear of one another in the governance practices of modern nations until quite late in the twentieth century. Elites including state officials, corporations, scientists, inventors, and financiers all participated in crafting the vast technological infrastructures that underpin modern societies,[37] with publics having little or no say in the desirability or not of particular directions of progress.

Three sets of policy initiatives in the past few decades responded to citizens' growing desire to reclaim the future from technological entrepreneurs and their corporate and governmental patrons. Each can claim early roots in U.S. policy, but each has also spread to other countries in ways that reveal national differences in public assessments of risks and benefits, as well as in underlying ideas of what citizens have a right to expect from governments. Each effort also ran into obstacles that demonstrate the difficulty of holding science and industry accountable to public values, especially when accountability means slowing down the advance of technology or, as in the case of GM crops, limiting the profitability of particular ways of redesigning nature-culture relationships. And although the United States was once the world leader in democratizing technological debates, the turn to privatization and market mechanisms in U.S. policy eroded that onetime position of primacy in all three policy domains.

Technology assessment, the most direct and formal means of forecasting and controlling sociotechnical futures, took wide hold across Western nations following the establishment of the U.S. Office of Technology Assessment. Although many of the OTA's offshoots and imitators proved more durable than their model, the OTA's own demise drew attention to some of the failings of its approach to making technology more democratic. Its assessment exercises were tied to the legislative process and suffered from some of the defects of lawmaking itself—captive to the politics of the present, dependent on uncertain public funding, and weakly responsive to the popular will or to rapid changes in circumstances. Constructive technology assessment looks more inclusive in principle and seeks to draw affected groups into holding technology accountable to users' perceived needs. But as the General Motors electric car and Desertec examples illustrate, CTA's utility clicks in too often, if at all, after the broad envisioning of societal futures has taken place somewhere outside the public's control. In neither case did the adoption or eventual abandonment of an envisioned technological future have much to do with the publics whose lives those projects would most directly have affected.

Ethics committees and public engagement exercises, though valuable for clarifying issues, also have shortcomings as mechanisms of democratic governance. The history of bioethics deliberation in the United States shows how profound moral issues became identified with routine forms of risk assessment and with utilitarian moves to keep the wheels of research turning while satisfying publics that ethical standards are being met. Bodies such as institutional review boards are not apt places for discussing the fundamental constitutional issues—including the very meaning of being human—that the genetic revolution raises. Bioethics has become one more specialist discourse that conveys a reassuring sense of democratic supervision while giving entrepreneurial scientific and technological imaginations free rein to determine in effect what counts as the public good.

Most paradoxically, even efforts to let publics inside the preserves of decisionmaking have proved only moderately successful in opening up entrenched traditions of decisionmaking.

GM Nation? offers one paradigm. In this case, arguably the most wide-ranging attempt ever undertaken by a national government to solicit public inputs on a particular technological pathway, disagreements arose about whether in the end the right public had expressed itself. In a political culture where consultation is typically invited by the state, British officials found it hard to swallow the sorts of "uninvited participation" that biotechnology critics saw as genuinely democratizing. Even America's more open administrative process, which in principle grants access to any interested or affected party wishing to have a voice in governing technology, has succumbed to the siren song of freedom to choose. By recasting technology as a source of consumer entitlements rather than as a form of politics, American policy has sacrificed the difficult discipline of deliberation for the easier path of informed consumerism. In the next and concluding chapter, we turn to today's vistas of technological advancement and the prospects they offer for deliberative, ethical, future making in a globally entangled but still deeply unequal world.

Chapter 9

INVENTION FOR THE PEOPLE

Through much of the twentieth century, the word "technology" conjured up the dirty, rusty, smelly, often invisible infrastructures of the first industrial revolution: "on tap but not on top," as Winston Churchill reportedly said of scientists advising governments. By the turn of the century, however, technology had come to be seen as a dynamic force for social change, bearer of ever larger hopes and fears as a global populace grew aware that technologies can transform, dramatically and perhaps irrevocably, the purposes and conditions of human existence. For the world population as a whole, the technological breakthroughs of the late twentieth century promised better health, swifter communication, cleaner environments, and unimaginable riches in knowledge and information. On the level of the individual human subject, the same technologies—powered by advances in nano-, bio-, info-, and cogno-sciences—created expansive opportunities for refashioning the self, offering hitherto unimaginable enhancements of life, along with fears that such powers might be misused or overused to humankind's lasting detriment.

The threat of nuclear annihilation was one powerful driver of a new global focus on technological risks, but from the 1980s onward technologies signaling what insiders called "convergent disruption" amplified public expectations about enhanced human capabilities while arousing new anxieties about technology's controlling and pervasive influence. Biotechnology promised miracles of life and health but awakened deep-seated concerns about environmental degradation and moral decay. More recently, developments in nanotechnology, synthetic biology, cognitive and neurosciences, and, above all, computing capability have focused attention on technology's immense potential to redefine the meaning of being human. Information technologies, in particular, herald an era in which the proliferating uses of personal data will render human subjects vulnerable to being mined and monitored as never before, and big data and algorithms may displace human discretion as apparently infallible instruments of governance. Meanwhile, climate change, a fearsome legacy of an older era of industrial production, hangs over the planet as a specter of what can happen when our appetite for technological advances is not reined in. Even in confronting this extreme threat, however, some remain confident that technology provides the best antidote for its own poisons, whether in the form of geoengineering to cool the Earth, new energy technologies to enable growth without pollution, or the ultimate science fiction fantasy of escaping to planets not wrecked by human mismanagement.

A trio of commonly held but flawed beliefs, each suggesting that technologies are fundamentally unmanageable, and therefore beyond ethical analysis and political supervision, long impeded systematic thinking about the governance of technology. The first is technological determinism, which holds, contrary to centuries of historical experience, that technologies have a built-in momentum that shapes and drives the course of history. The second is technocracy, which implies that only skilled and knowledgeable experts possess the competence to govern the advance of technology. The third is unintended consequences, a notion that implicitly positions the harms caused by technology outside the ambit of intention or forethought, and thus breeds fatalism about the very possibility of bringing our nonhuman, mechanical, or cognitive collaborators under meaningful human control.

We have seen throughout this book that technological systems are in fact more plastic and more amenable to ethical and political oversight than these strands of conventional wisdom hold. The making and deploying of technologies have given rise to ethical questions on multiple levels, from how to protect individual values and beliefs to how much respect to accord to the policy intuitions of nation-states informed by distinctive legal and political cultures. These questions gained urgency through the global spread of technology in the later twentieth century, and they impose on technologically advanced societies an obligation of active public reflection and response. Pharmaceutical drugs, for example, were once seen simply as miraculous breakthroughs in medical knowledge and practice, bringing benefits to all humanity. In the global marketplace, however, drugs have become entangled with a host of normative issues bearing not just on national but on transnational rights and duties. These include ownership of human biological materials, privacy of medical data, consent to clinical trials, ethics of human subjects research, manufacture of generic drugs, access to experimental and essential medicines, protection of indigenous knowledge, and many related matters. On all of these issues, stakes and expectations vary widely, depending on the observer's socioeconomic position—sick or well, rich or poor, producer or consumer, and citizen of a developed or a developing country. Pharmaceuticals thus exert power far beyond their immediate properties as therapeutic agents: they reconfigure the human body in transnational space, as an object of science, medicine, economics, law, and policy. Yet processes of deliberation on the politics and ethics of drug development remain significantly less available at the supranational than at the national level.

How can our far-reaching technological inventions be governed so that they meet the ethical needs of a globalizing world? Who should assess the risks and benefits of innovation, especially when the results cut across national boundaries: according to whose criteria, in consultation with which affected groups, subject to what procedural safeguards, and with what remedies if decisions prove misguided or injurious? Technological mishaps and missteps continually alert us to the need for deeper social, political, and legal analysis. Yet many basic issues of right and wrong remain deeply contested, and principles for resolving them glimmer at best weakly on the horizon. This concluding chapter reviews the major insights gleaned from decades of national and global experience with the governance of technological systems, as discussed and exemplified in earlier chapters. The challenges that still lie ahead are grouped under three thematic headings that tie those histories together: anticipation, ownership, and responsibility.

ANTICIPATION: AN UNEQUAL GIFT

If there is one thread that runs through all policy discourses on technology, it is the need for wise anticipation. The technology assessment programs of the 1970s and 1980s grew out of an interest in foreseeing and forestalling technologically induced physical and environmental harms. Similarly, programs to evaluate the ethical implications of new technologies reflect a widely distributed desire to anticipate and ward off insupportable moral harm. Unsurprisingly perhaps, in modern societies anticipation of both physical and moral consequences is closely tied to expert predictions of likely outcomes. The linear model of technology policy, which begins with risk assessment and only later injects values into decisionmaking, delegates the initial phase of anticipation to experts, or technocrats, who are thought to understand best how technology operates, and where it may go wrong. Yet, as the stories scattered through the preceding chapters illustrate, experts' imaginations are often circumscribed by the very nature of their expertise. The known takes precedence over the unknown. Accordingly, anticipation by experts tends to foreground short-term, calculable, and uncontested effects over those deemed speculative, far-fetched, or politically contentious. And, drunk with the mastery of pathbreaking scientific insights, experts too often underestimate the complexities of hybrid sociotechnical systems, in which ill-understood dynamics and feedbacks between human and nonhuman elements confound the precision of lab-based expectations.

Who Imagines the Future?

In the early, promissory stages of technological change, foresight is often shaped and limited by the disciplinary competence of expert advocates. The molecular biologists at Asilomar were determined to prevent a biological catastrophe, but they did not imagine a world reshaped by commercial biotechnology in which major GM crops would routinely displace their wild counterparts, even though that world came into being in America within decades of the discovery of recombinant DNA technology. The German engineering companies that conceived the transfer of solar energy from the Sahara to

Europe did not think about the political management of such an audacious transnational project, spanning vastly different cultures and economies, and Desertec eventually fell victim to the ungovernability of its designers' imagination.

Institutional conservatism also precludes farsighted prediction. Courts, for example, seek to ensure stability in the law, but at the cost of ignoring the values affected by technological transformations. Genentech's characterization of Ananda Chakrabarty's bacterium as mere matter won out in a common-law high court that preferred the biotech company's incremental vision over Jeremy Rifkin's seemingly unfounded concerns about a slippery slope from the patenting of bacteria to the patenting of higher animals. In hindsight, however, it seems that Rifkin more accurately foresaw the likely trajectory of patenting life than Genentech and its prominent advocates in the scientific community. The Canadian Supreme Court's rejection of a patent on the oncomouse and the U.S. Supreme Court's own retreat from patenting human genes display a retroactive (some might say belated) repudiation of the more extreme implications of patenting life.

In expert thinking, too, there is frequently a tacit slippage between is and ought that dulls the edge of ethical concern. Any departure from the common sense of scientists is deemed unreasonable, fictional, or fantastic, and what cannot (yet) be done is not considered worth worrying about. Invention, as we have seen, tends to be regarded as a good in itself, with ethical oversight invoked chiefly to ensure that the promised good will not be derailed through a reckless greed for profit. Predictions of technical improbability then serve as a barrier against potentially "unrealistic" ethical speculation and premature public anxiety. Consider, for example, a brief entry from *Science*'s 2013 list of runners-up for breakthrough of the year. Under the heading "Human Cloning at Last," the journal reported, "This year, researchers announced they had cloned human embryos and used them as a source of embryonic stem (ES) cells—a long cherished goal." The report went on to downplay the ethical meaning of this development, citing improbability rather than moral aversion as the reason why society should not worry: "The feat also raises concerns about cloned babies. *But that seems unlikely for now.* Despite hundreds of tries, the Oregon researchers say, none of their cloned monkey embryos have established a pregnancy in surrogate females" (emphasis added).[1] Silenced in this account is the what-if question. Suppose a pregnancy had been established and cloned babies made to seem more probable. Is it fitting that societies of such infinitely creative capacity as ours should reflect on the ethical implications of such far-reaching technological experiments only after a threat to human dignity comes knocking at the door?

In the world of innovation imagined by the *Science* report, one wonders when, if ever, it would be appropriate for the public to voice concerns about cloned babies. Only after researchers in Oregon or elsewhere announced they had succeeded in inducing a pregnancy with their cloned embryos? Surely, though, by that stage a new "is" would overwhelm the possibility of a more nuanced and precautionary "ought"—such as a rule that might have made researchers think twice, or think aloud in public, before making the "hundreds of tries" to induce an ethically controversial pregnancy. Further, the description of human embryo cloning as "a long cherished goal" overlooks the contested and divergent moral settlements that exist across nations about where an embryo ends and a baby begins. In Germany, for example, concern might arise further upstream in the research process than in the United States, illustrating different preferences for deontological as opposed to utilitarian arguments in two nations whose ethical thinking is, in both cases, rooted in the "Western" tradition. Given such disparities in basic commitments to ethical reasoning, is it even appropriate for decisions of such consequence for the entire human species to be undertaken by technologically leading nations going it alone?

With confident expert assertions as a baseline for moral analysis, ethical assessments tend in any case to devolve into utilitarian analyses of the costs and benefits of probable scenarios. This is exactly what happened when two different U.S. presidential ethics commissions evaluated human cloning and synthetic biology, respectively. In each case, the commission concluded that there were no ethical worries at present because the technology was not yet sufficiently safe to use.[2] Principled questions about which kinds of worlds, containing what forms of life, should be made with technology got sidelined in favor of shallower assessments of imminent risk. Such disappointingly constrained exercises contradict the hopes of some democracy theorists that anticipatory governance, enacted through procedures such as constructive technology assessment (see chapter 8), "can contribute to bending the long arc of technoscience more toward humane ends."[3]

Trickle-Down Innovation

Modern societies have committed extensive resources of money and expertise to anticipating adverse technological futures and warding off potential ill effects, but those resources are unevenly distributed across nations and technological domains. Events like the 2013 Rana Plaza factory collapse in Bangladesh speak to a global political economy in which the lives of the poor are placed at risk in ways that would be deemed intolerable in richer countries. Astonishingly, responsibility proved almost as hard to pin down in 2013, after the textile industry's worst-ever disaster, as it had been nearly thirty years earlier, in 1984, when the Union Carbide plant in Bhopal released a deadly cloud of methyl isocyanate over a sleeping, unsuspecting city. It took more than two years for charges against Sohel Rana to be formalized. Although many clothing companies employing Bangladeshi labor pledged contributions to a compensation fund, there was no global forum in which lines of responsibility could be formally established or major transnational actors be publicly held to account. To the extent that risk assessment is meant to foresee and prevent such horrific accidents, it has not yet caught up with the challenges of prediction and protection under conditions of severe global economic, political, and informational inequality.

The failures of risk assessment can also be attributed to the narrow causal frames that have historically underpinned this widely used anticipatory technique. Construed as a "science," risk assessment perennially downplays social factors and overemphasizes variables that can be quantified as against more elusive economic, institutional, and cultural contributions to risk creation. It took two space shuttle disasters, the loss of the *Columbia* as well as the *Challenger*, for U.S. accident experts to begin identifying organizational conditions within NASA that made these tragedies possible. The cataclysmic Bhopal disaster produced no comparable moment of reckoning. It led to what many still see as a premature and unjust settlement, precluding a fair adjudication of all that went wrong when a highly hazardous technology was transferred between nations lacking parity in wealth and expertise.

Despite its limitations as an instrument of governance, anticipation is a value no society would care to live without. Throughout history, human beings have turned to anticipated futures—whether in this life or one after—as a means of offsetting the terrors and trials of the present. Visions of a better world have driven inventors to imagine and create new tools, from the steam engine, cotton gin, and lightbulb to today's genetically modified Innate potato, which is alleged to produce less cancer-causing acrylamide when fried.[4] Visions of emancipation fire the California-based Singularity University's ambition to improve lives on an exponential scale, one billion people at a time.[5] Anticipation in this optimistic sense also powers the economics of innovation. Investors buy shares in fledgling technology companies in anticipation of elevated profits somewhere down the long road of research and development. Consumers, too, do their part, as when they hang on Apple's latest moves, anticipating unimagined gains in sleekness, versatility, and speed. One of the blessings of modernity is that the gap between idea and actualization has been shortened. Our ability to translate knowledge into inventions has grown along with our scientific and economic resources. There has never been a time in human history when bright ideas were more likely to attract quick attention, venture capital, and the legal and political support needed to bring them into global markets, in usable form and in real time.

As yet, though, the luxury of positive anticipation is limited largely to those who already have much, and hence are well positioned to dream up what it would mean to have still more. Steve Jobs's famous claim that he knew what consumers want before they themselves did was formulated in an economy of plenty, not want; Jobs presumed that people would not only want but also have the resources to buy the beautiful gadgets he dreamed up for them. But most of the world's masses are in no position to anticipate for themselves either immediate benefits or improved long-term prospects from the forward march of technology. They must accept the promise of benevolent outsiders that their lives will be bettered through inventions designed elsewhere, by entrepreneurs closer to technology's moving frontiers, with the capital and know-how to engineer large-scale change. Inequality—not only of access but even more of anticipation—thus emerges as an unresolved ethical and political barrier to the just governance of technological innovation.

One idea that has gained ground to offset the unfair imaginative advantages enjoyed by technologically and economically better endowed societies is "frugal innovation" or "frugal engineering." Frugality here means making technologies that are better adapted to the means of the less well-off. A poster child is the $100 laptop project initiated by Nicolas Negroponte of the Massachusetts Institute of Technology and embraced by former UN Secretary-General Kofi Annan in 2005. The goal was to create an easily affordable computer for worldwide distribution by stripping away unnecessary frills and equipping it to work under unfavorable conditions such as the erratic power supply that plagues most developing nations. Frugal innovations embrace everything from the Indian Tata Group's Nano, the cheapest car in the world, to the Nokia 1100, a popular, stripped-down cell phone, and Unilever's single-use toiletries. One could even see the rise of the "shared economy"—with its transformation of "extra" rooms, cars, and even pets into spare capital—as a kind of frugal innovation potentially using the spare goods of the haves to benefit the masses of have-nots.

Although they gained ground in the economics of plenty (where else is there so much stuff to spare?), shared economies arguably bear a family resemblance to Muhammad Yunus's influential notion of microcredit, which also drew on the principle that one need not have much to be able to share and thereby produce more profit for all. Additionally, certain material innovations for the poor, such as smokeless cooking stoves, composting toilets, and simple water purification systems, have drawn the attention of Northern engineers and scientists and could prove hugely beneficial. On the whole, however, the gap between the subsistence innovations aimed at the poor and the kinds of visionary ideas that propel breakthrough developments in the convergent technologies of wealthy societies remains as staggering as ever.

Laudable as the "no frills" commodities are, they still largely represent a kind of trickle-down theory of innovation, in which the technological achievements of the wealthy and well resourced define the anticipatory horizons of the less privileged. What the rich invented to fit their circumstances remains the gold standard for what the poor should need and want, only with fewer features, less sensory appeal, and possibly less likelihood of serving as platforms for autonomous development.[6] The problem of innovation is rarely posed in reverse. What kinds of technological futures would make most sense for the two-fifths or so of the world's population who subsist on less than two dollars a day?[7] Should they spend fifty days of scarce income on a kid's laptop, or are there more pressing demands that technology could more advantageously correct? In any case, would having the laptop open doors to the sorts of playful yet productive connections (Facebook, Reddit, Pinterest, Instagram) that the rich can effortlessly imagine and indulge in?

The power differential that permeates the design of technological futures is worrisome, but that is not humanity's only ethical concern. Indeed, in an era when we are more than ever conscious of the unsustainability of high-consuming lifestyles, it is unclear that the futures envisioned by the rich should take precedence over the imaginations of the poor. As downsizing, greening, simplifying, and even "freeganism" (living from others' unwanted trash) rise in appeal among the young, better ideas for how to live more lightly on Earth may well need to come from below, ideas that pay more attention to the values of social cohesion and stewardship than to the ceaseless imperialist exploitation of living and nonliving materials.

OWNERSHIP AND INVENTION

Technological development over the past several centuries owes as much to ideas of property, or private gain, as to anticipations of the collective good. New technologies as often as not involve new modes of extraction, generating huge profits for those who discover how to access and distribute previously unappropriated resources. Prospecting and mining are obvious examples, but technology has enabled many other forms of resource use, often linked to projects of state and imperial expansion. California became a modern state in the mid-nineteenth century through an influx of hundreds of thousands of immigrants staking claims to newfound gold. Diamond mines supported Cecil Rhodes's imperial ambitions in South Africa, and the rubber and ivory trade underpinned King Leopold II's brutal regime in the Congo. Exploitation, cruelty, and misuse of power overshadowed these enterprises, but so long as demand existed for the commodities they generated, technology, economics, and politics combined to keep oppressive regimes in place.

Whereas the earliest extractive technologies pulled things that people already valued out of rocks, earth, plants, and oceans, many technologies of today assign value to things that were not historically treated as commodities. Biological materials, everything from genes to novel lab-created entities, such as the cancer-prone Harvard oncomouse, belong in this category. Constructs such as carbon markets and ecosystem services have turned parts of nature into tradable commodities, although one cannot own or use them in the way one uses gold. Also in this category are the burgeoning compilations of big data produced through the information revolution.[8] Social media have commodified people's habits and preferences, their memories and their aspirations, through large-scale aggregation. Facebook trades on the fact that more than a billion people wish to be connected to their "friends" and are willing to contribute masses of personal information to a private enterprise in return for reaching those friends and knowing what they are up to. Twitter capitalizes on people's passing thoughts in 140 characters and attached images that few would have seen fit to share until the Internet enabled fleeting impressions, mental or visual, to be disseminated to global audiences. Pinterest thrives on people sharing dreams of weddings, vacations, home renovations, and other things people want to welcome into their anticipated futures. Direct-to-consumer testing companies compile databases of voluntarily supplied genetic information that have commercial potential for pharmaceutical research and development. All of these technological systems profit from people in novel ways, mining their thoughts, words, habits, bodies, and emotions as resources to create new marketable goods.

The tortuous story of the HeLa cell line demonstrated that people's sense of control over their own bodies—and perhaps even more their minds—can diverge significantly from the presumptions of ownership built into existing law and policy. Rebecca Skloot's prizewinning reconstruction of Henrietta Lacks's story ignited a volatile mix of one of modern biology's most useful research tools and America's perennially vexed narratives of race and poverty. The National Institutes of Health recognized a public relations disaster in the making if the moral claims of a historically excluded group, intimately bound up with past failures of biomedical ethics, were left untended. A onetime procedure solved the NIH's immediate problem, granting the Lacks family the right to participate in decisions involving their ancestor's biological legacy. Shorn of its unique historical and political trappings, the case of the HeLa cell line would likely have foundered, leaving Henrietta Lacks to die a second death. That extraordinary example is unlikely to serve as a persuasive precedent for situations in which science and its objects of study enjoy less symmetrical bargaining positions.

For the most part, the intellectual property regimes that govern technological innovation continue to uphold ideas of ownership and capital that originated in the modern industrial world some two hundred years ago. These ideas are not completely homogeneous, of course, even across Western nations. European patent law differs from its American equivalent in providing an explicit mandate against inventions that violate *ordre public*, or morality. European law also sets higher bars for demonstrating what constitutes an "inventive step" than U.S. law in its most permissive phases, as when the Patent and Trademark Office for a time granted patents on isolated DNA fragments of no demonstrated utility. On the whole, however, intellectual property hides its normative foundations—such as favoring individual entrepreneurship over collective effort— beneath a veil of technical neutrality. In the Gleevec case (see chapter 7), the Indian Supreme Court explicitly stated that patent law is an instrument of economic development, so that the scope and nature of the protections it offers should reflect "the economic conditions of the country."[9] Such language underscoring the normative foundations and power asymmetries of intellectual property rights is rarely encountered in Western legal decisions.

Yet the law can also confirm or destabilize ownership claims whose entanglement with public concerns about autonomy, privacy, life, or health have become visible and undeniable. Legal power emerged as salient in the U.S. Supreme Court's 2013 decision to discontinue the patenting of human genes, overriding the Patent and Trademark Office's policy that had held otherwise for decades. Years of activism by a public interest group, the American Civil Liberties Union, undid a commodifying move that violated tacit but widely held public values about the rightfulness of privatizing parts of the human genome. Such about-faces can occur even in the international arena, although here the driver tends to be political economy rather than the politics of personhood. Thus, the Convention on Biological Diversity sought to reverse centuries of unregulated biopiracy by recognizing the ownership claims of indigenous knowledge holders. The treaty provides for profit sharing between local communities and bioprospecting companies that develop new therapeutic compounds by using local knowledge and materials. Economic disparities in the distribution rather than the production of drugs motivated the 2001 Doha Declaration, amending the Agreement on Trade-Related Aspects of Intellectual Property Rights (the TRIPS Agreement). TRIPS now allows countries to circumvent pharmaceutical patent rights in emergency situations such as the AIDS crisis that require rapid, expanded access to essential medicines. Unexamined questions for the future include how to recognize distributed ownership of intellectual property as invention, increasingly, relies on the circulation and synergy of people, ideas, and materials across formerly unbridgeable geopolitical boundaries.

RESPONSIBILITY: PUBLIC AND PRIVATE

Technological advances can remake the divisions between public and private spaces in ways that affect not only personal autonomy and opportunities for public deliberation but also, and perhaps more significantly, the norms of individual and collective responsibility. From Karl Marx to Herbert Marcuse, social theorists have warned of the flattening, standardizing, and dulling effects of technology on the human spirit, through the spread of industrial work practices and mass culture. Inside the iron cages of large technological systems, subject to the tyranny of the assembly line and the production quota, emancipation and responsibility for the self may sound like cosmic jokes. New biological and informational technologies carry big promises of liberation, especially from inherited disease; but they also allow unprecedented access into bodies and minds and create possibilities for social control that surpass even the Orwellian nightmare of *Nineteen Eighty-Four.*

To some degree, people themselves have proved complicit in shrinking the boundaries of the private in the digital age. The rise of the culture of the selfie (self-portraits taken with handheld cameras) and people's seemingly limitless appetite for attracting the gaze of others have created an intoxicating space in which indiscretion can wreck employment prospects, ruin political careers, and intrude into zones that even celebrities once kept to themselves.[10] As murky as they are pervasive, data sets have emerged as a prime source of contestation in the struggle between individual self-control and overweening state and corporate power. Data oligarchs currently operate under a patchwork of state-sponsored and self-imposed regulation, such as Google's and Facebook's privacy policies, reflecting disparate normative commitments. To bring information's wild frontiers under anything approaching systematic control may require wider adoption of concepts such as the European Union's "data subject," as well as sustained debate on what people know, expect, and will tolerate with regard to the circulation of data about their preferences and their persons.

A different kind of ethical worry stems from the retrenchment of the very idea of "public-ness" in an era dominated by market thinking and neoliberal forms of governance. Legislatures, classic seats of representative democracy, seem increasingly less relevant in a time when big money can exert pressure on public opinion through a wide diversity of channels. Entrepreneurs fiercely resist legislation, arguing (particularly in the United States) that it stifles innovation because law constantly lags behind science and technology. Charismatic figures such as Sir Timothy Berners-Lee, inventor of the World Wide Web, and Craig Venter, co-sequencer of the human genome, point to the success of the Internet and the spread of biotechnology as prime examples of the benefits of laissez-faire technological development. Legislatures for their part often lack courage and expertise to fight back, and may be politically beholden to the very interests that they are theoretically committed to keeping under control.

The explosion of ethics bodies dedicated to specific technological developments—biomedicine, nanotechnology, synthetic biology, neurosciences—should offer relief against private capture of public policymaking institutions, but in practice these often invisible committees add to the layers of ethical concern. When tied too closely to the research enterprise, bodies such as institutional review boards supervising human subjects research tend to operate with a tacit commitment not to burden their home institutions or their scientific stars with too many demands. As we have seen, even the more visible national ethics commissions prefer the safe haven of cost-benefit analysis of imminent technological futures to asking hard questions about the longer-term public benefits of new lines of research. Britain's Warnock Commission, for example, performed a huge service for modern biomedicine by characterizing the pre-fourteen-day embryo as a nonhuman for research purposes. In Britain, that bright line created a permissive space for frontiers research using embryonic tissues and led to several first-time approvals for new ethical extensions, as in the use of "savior siblings" to provide matching tissue for an ill child or the making of the three-person embryo to eliminate the mother's mitochondrial genetic disease. Yet, left unchecked, this relative privatization of public morality inside closed-loop ethics bodies may spawn public alienation and normative rejection. Ongoing U.S. controversies over issues such as stem cell research reveal the political vulnerability of prescriptive line-drawing by ethical experts whose views are not exposed to ongoing public scrutiny and reapproval.

CONCLUSION

Modern societies, in the past hundred years, have invented many strange and wonderful things and broken unimagined barriers in mobility, communication, calculation, and preservation of life and health. With technology, we have turned famine to surplus, eliminated killer diseases, plumbed the oceans and the stratosphere, brought outer space within the horizons of human imagination, and opened up the recesses of the human mind to targeted exploration. Much of the world rejoices when a robot lands on Mars or on a comet, in a feat of accuracy that some say is like throwing a hammer from London and hitting a nail in New Delhi. The dream of constructing reusable rockets became a reality in December 2015 when a company headed by the U.S. technology entrepreneur Elon Musk successfully landed a rocket back on Earth after launching satellites into orbit. Concurrently, huge strides have been made in monitoring, modeling, and measuring the impact of technological activities on health, environment, and society. Technological cultures are no longer so heedless of nature as when nineteenth-century factories belched smoke and acrid dust over pristine landscapes, toxic dyestuffs ran into rivers uncontrolled, or companies hid evidence that workers exposed to asbestos were dying of lung disease in the thousands.

Yet, as we have seen throughout the preceding chapters, institutional deficiencies, unequal resources, and complacent storytelling continue to hamper profound reflection on the intersections and mutual influences of technology and human values. Important perspectives that might favor caution or precaution tend to be shunted aside in what feels at times like a heedless rush toward the new. As a result, the potential that technology holds for emancipation, creativity, and empowerment remains unfulfilled or at best woefully ill distributed. Issues that cry out for careful forethought and sustained global attention, such as the genomic and information revolutions, are depoliticized or rendered invisible by opportunistic design choices whose partially path-dependent tracks frustrate future creativity and liberation.

The life history of the automobile, one of the twentieth century's most successful technological inventions, and still a fetish of wealth the world over, remains a paradigmatic case study in the limits of human foresight. The car unlocked immense possibilities for individual freedom and productivity, but these went hand in hand with drastic consequences for society that no one had imagined or regulated in timely fashion: more than a million traffic deaths worldwide each year, the spread of deadening, routinized work practices, the blight of urban air pollution, the fragmentation of communities, the decay of once-great manufacturing centers, and eventually world-threatening climate change. Could current practices of responsible innovation and anticipatory governance have turned the tide of the automobile's history before it took a tragic course? For technologies of mass appeal and enormous economic and social consequence, the localized and episodic processes of governance commanded by nation-states seem sadly inadequate. Occasional mobilization, moreover, fails to get to the heart of the asymmetries of anticipation. For all practical purposes, the power to set the rules of the game for governing technology rests with capital and industry, not with the political representatives of the working, consuming, and too often suffering masses.

This deep democratic deficit cannot be cured with procedural Band-Aids. The recently proliferating experiments with public consultation, constructive technology assessment, and ethical review do no harm and should certainly continue. They have the merit of keeping people involved in decisions pertaining to their everyday lives, and over time they may clarify a society's preferences for rule by technology. Such ad hoc processes, however, are not a substitute for the kind of constitutional convention that our grand bargain with technology in effect demands. To unleash the potential of the democratic imagination, contemporary societies will have to acknowledge first of all that technology is neither self-propelling nor value-free. Even modest technological improvements create new normative rights and obligations, as when I have to cross the street in Cambridge under the watchful eye of traffic lights at an intersection that once was unregulated. The parallels between technology and law then become apparent, showing that the former no less than the latter is a potent instrument for fashioning our collective futures. That recognition should spur a deeper ethical and political engagement in the governance of technology. Only if we acknowledge technology's power to shape our hearts and minds, and our collective beliefs and behaviors, will the discourses of governance shift from fatalistic determinism to the emancipation of self-determination. Only then will an ethic of equal rights of anticipation be accepted as foundational to human civilization on our fragile and burdened planet.

ACKNOWLEDGMENTS

A book is the end of a journey, and good journeys happen in company. This book was on its way longer than originally planned. It began life as a simple monograph on the risks of modern technology, but, as the publisher's vision for the series changed, it turned into a more complex reflection on political inclusion and exclusion in the crafting of humanity's technological future. I would like to thank the editorial team at Norton for its steady support throughout that transformation. Anthony Appiah set things in motion by urging me to take on the project in the first place. Roby Harrington offered much appreciated patience during downtimes, and Brendan Curry's smart and careful reading helped clarify the main messages of a manuscript that sometimes got lost in expository details. I am also grateful to Sophie Duvernoy and Nathaniel Dennett for smoothly shepherding the book through the logistics of editing and publication.

Some empirical material covered in the book represents new research on law and technology; much of the rest derives from earlier work revisited in a new analytic context. As a result, the book owes less to particular individuals or projects than it does to collective ways of thinking and working fostered in the

Program on Science and Technology Studies that I direct at the Harvard Kennedy School. Weekly meetings with fellows in the program over the past few years offered a space for deepening and honing my ideas on the ethical and political dimensions of governing technology. Three conversation partners deserve mention by name: Rob Hagendijk, Ben Hurlbut, and Hilton Simmet. Much of what this book says about the constitutive—and constitutional—role of technology and democracy in modernity reflects my ongoing exchanges with each of them, though none bears any responsibility for omissions and weaknesses that are, as always, the author's own.

I also benefited enormously from working, as principal investigator, on two National Science Foundation (NSF) grants related to the book's subject matter: "The Fukushima Disaster and the Politics of Nuclear Power in the United States and Japan" (NSF Award No. 1257117) and "Traveling Imaginaries of Innovation: The Practice Turn and Its Transnational Implementation" (NSF Award No. 1457011). NSF has played a crucial role in the evolution of my career, as well as my efforts to build the field of science and technology studies, and I am pleased once again to acknowledge the Foundation's support. I would also like to thank Shana Rabinowich, whose contributions to the successful management of all my projects for the past six years cannot be overstated.

Lastly, my family's presence in my life is too foundational to need many words of acknowledgement. Since this is a book about the future, it seems especially appropriate to dedicate it to Nina, who will have a hand in shaping a world that I can at best dimly imagine.

NOTES

Chapter 1: The Power of Technology

1. When Facebook crossed the 1.3 billion member mark in January 2015, it began to be called the largest country in the world. Comparisons of Facebook to countries had started years earlier. See Steven Mostyn, "Facebook Population Equivalent to Third-Biggest Country on Earth," *Tech Herald*, July 22, 2010.
2. WHO, World Health Report, "50 Facts: Global Health Situation and Trends 1955–2025," http://www.who.int/whr/1998/media_centre/50facts/en/, accessed October 2015.
3. Nick Bostrom, "Existential Risks: Analyzing Human Extinction Scenarios and Related Hazards," *Journal of Evolution and Technology* 9, no. 1 (2002).
4. Richard Rhodes, *Deadly Feasts: The "Prion" Controversy and the Public's Health* (New York: Simon and Schuster, 1997).
5. United Nations, Economic and Social Council, *World Mortality Report* 2013 (New York: United Nations, 2013).
6. World Health Organization, Global Health Observatory (GHO) data, http://www.who.int/gho/child_health/mortality/neonatal_infant_text/en/, accessed October 2015.
7. See http://www.worldpopulationbalance.org/energy_india, accessed October 2015.
8. Lee Raine and D'Vera Cohen, "Census: Computer Ownership, Internet Connection Varies Widely across U.S.," Pew Research Center, September 19, 2014, http://www.pewresearch.org/fact-tank/2014/09/19/census-computer-ownership-internet-connection-varies-widely-across-u-s/, accessed October 2015.

9. The Dark Mountain Project, FAQs, http://dark-mountain.net/about/faqs/, accessed December 2015.

10. Robert D. Putnam, *Bowling Alone: The Collapse and Revival of American Community* (New York: Simon and Schuster, 2000).

11. Sherry Turkle, *Alone Together: Why We Expect More from Technology and Less from Each Other* (New York: Basic Books, 2011).

12. Leo Marx, "Technology: The Emergence of a Hazardous Concept," *Technology and Culture* 51, no. 3 (2010): 561–77.

13. Michael Pollan, *The Omnivore's Dilemma: A Natural History of Four Meals* (New York: Penguin, 2006).

14. Arthur C. Clarke, *2001: A Space Odyssey* (New York: New American Library, 1968).

15. Bill Joy, "Why the Future Doesn't Need Us," *Wired*, April 1, 2000, http://www.wired.com/wired/archive/8.04/joy.html?pg=3&topic=&topic_set=.

16. Ibid.

17. Clay McShane, "The Origins and Globalization of Traffic Control Signals," *Journal of Urban History* 25 (1999): 379–404.

18. Tom McNichol, "Roads Gone Wild," *Wired*, December 1, 2004, http://www.wired.com/wired/archive/12.12/traffic.html.

19. Marx, "Technology," p. 564.

20. Langdon Winner, "Do Artifacts Have Politics?," *Daedalus* 109, no. 1 (1980): 121–36.

21. Claude Henri de Saint-Simon, *Political Thought of Saint-Simon*, ed. Ghita Ionescu, trans. Valence Ionescu (Oxford: Oxford University Press, 1976).

22. Sheila Jasanoff, *The Fifth Branch: Science Advisers as Policymakers* (Cambridge, MA: Harvard University Press, 1990).

23. Harold J. Laski, *The Limitations of the Expert* (London: Fabian Society, 1931), p. 4.

24. *Buck v. Bell*, 274 U.S. 200 (1927).

25. Diane Vaughan, *The Challenger Launch Decision: Risky Technology, Culture, and Deviance at NASA* (Chicago: University of Chicago Press, 1996).

26. Charles Ferguson, "Larry Summers and the Subversion of Economics," *Chronicle of Higher Education*, October 3, 2010, http://chronicle.com/article/Larry-Summersthe/124790/, accessed October 2015.

27. Apple iPhone 4 press conference, July 16, 2010, Cupertino, CA, http://www.apple.com/apple-events/july-2010/, accessed February 2012.

28. Michael Rose, "'Antennagate' Press Conference Video and Official Pages Up," *Engadget*, July 16, 2010, http://www.tuaw.com/2010/07/16/

antennagate-press-conference-video-and-web-pages-up/, accessed February 2012.

29. Robert J. Lopez and Rich Connell, "Metrolink Engineer Let Unauthorized 'Rail Enthusiasts' Control Train," *Los Angeles Times*, March 3, 2009.

30. Sheila Jasanoff, *Designs on Nature: Science and Democracy in Europe and the United States* (Princeton, NJ: Princeton University Press, 2005).

31. On the eve of the 2015 climate talks in Paris, the tech billionaire Peter Thiel authored an op-ed entitled "The New Atomic Age We Need," *New York Times*, November 27, 2015.

Chapter 2: Risk and Responsibility

1. Wiebe E. Bijker, Thomas P. Hughes, and Trevor J. Pinch, eds., *The Social Construction of Technological Systems: New Directions in the Sociology and History of Technology* (Cambridge, MA: MIT Press, 1987).

2. Coral Davenport and Jack Ewing, "VW Is Said to Cheat on Diesel Emissions; U.S. to Order Big Recall," *New York Times*, September 18, 2015.

3. Paul J. Crutzen and Eugene F. Stoermer, "The 'Anthropocene,'" *Global Change Newsletter*, no. 41 (May 2000): 17–18, available at, http://www.igbp.net/download/18.316f18321323470177580001401/1376383088452/NL41.pdf, accessed November 2015.

4. Ulrich Beck, *Risk Society: Towards a New Modernity*, trans. Mark Ritter (London: Sage, 1992).

5. Bill Vlasic, "G.M. Inquiry Cites Years of Neglect over Fatal Defect," *New York Times*, June 5, 2014. General Motors eventually agreed to pay $900 million to settle criminal charges stemming from the episode.

6. Charles Perrow, *Normal Accidents: Living with High-Risk Technologies* (New York: Basic Books, 1984).

7. Rogers Commission Report, Report of the Presidential Commission on the Space Shuttle Challenger Accident (1986); Columbia Accident Investigation Board, Report (2003).

8. Diane Vaughan, *The Challenger Launch Decision: Risky Technology, Culture, and Deviance at NASA* (Chicago: University of Chicago Press, 1996).

9. Ronald Brickman, Sheila Jasanoff and Thomas Ilgen, *Controlling Chemicals: The Politics of Regulation in Europe and the United States* (Ithaca, NY: Cornell University Press, 1985).

10. National Research Council, *Risk Assessment in the Federal Government: Managing the Process* (Washington, DC: National Academies Press, 1983).

11. Erving Goffman, *Frame Analysis: An Essay on the Organization of Experience* (New York: Harper and Row, 1974).

12. Dorothy Nelkin, *Nuclear Power and Its Critics: The Cayuga Lake Controversy* (Ithaca, NY: Cornell University Press, 1971).

13. Sheldon Krimsky, *Genetic Alchemy: The Social History of the Recombinant DNA Controversy* (Cambridge, MA: MIT Press, 1982).

14. Sheila Jasanoff, *The Fifth Branch: Science Advisers as Policymakers* (Cambridge, MA: Harvard University Press, 1990).

15. Jason Corburn, *Street Science: Community Knowledge and Environmental Health Justice* (Cambridge, MA: MIT Press, 2005).

Chapter 3: The Ethical Anatomy of Disasters

1. Typically, such numbers are never completely certain. Reported numbers of the dead in the Rana Plaza collapse vary from 1,127 to 1,131.

2. For an exceptionally well-researched account of the disaster, see Hauke Goos and Ralf Hoppe, "Made in Bangladesh: Greed, Globalization and the Dhaka Tragedy," *Der Spiegel*, July 1, 2013 (English translation by Christopher Sultan). See also Jim Yardley, "The Most Hated Bangladeshi, Toppled from a Shady Empire," *New York Times*, April 30, 2013.

3. European Food Safety Authority, "Shiga Toxin-Producing *E. coli* (STEC) O104:H4 2011 Outbreaks in Europe: Taking Stock," *EFSA Journal*, October 3, 2011.

4. "2011 Outbreak of Rare E. Coli Strain Was Costly for Europe," *Food Safety News*, April 3, 2015, http://www.foodsafetynews.com/2015/04/2011-outbreak-of-rare-e-coli-strain-was-costly-for-europe/#.Vm8Wfb8xRAM.

5. European Food Safety Authority, "Shiga Toxin-Producing *E. coli*."

6. James Kanter, "Death Toll Rises in E. coli Outbreak," *New York Times*, May 31, 2011.

7. Dozens of books have been written about the Bhopal disaster from multiple points of view. A small sampling includes Upendra Baxi and Amita Dhanda, *Valiant Victims and Lethal Litigation: The Bhopal Case* (Delhi: Indian Law Institute, 1990); Kim Fortun, *Advocacy after Bhopal: Environmentalism, Disaster, New Global Orders* (Chicago: University of Chicago Press, 2001); and Dominique Lapierre and Javier Moro, *Five Past Midnight: The Epic Story of the World's Deadliest Industrial Disaster* (New York: Warner, 2002).

8. Robert D. McFadden, "India Disaster: Chronicle of a Nightmare," *New York Times*, December 10, 1984.

9. http://www.unioncarbide.com/History.

10. Friends of the Earth Malaysia, *The Bhopal Tragedy—One Year After* (Penang: APPEN, 1985), p. 44.

11. Environmental Working Group, Chemical Industry Archives, "The Inside Story: Bhopal," http://www.chemicalindustryarchives.org/dirty-secrets/bhopal/index.asp.

12. *In re Union Carbide Corporation Gas Plant Disaster at Bhopal, India in December, 1984,* 634 F. Supp. 842 (1986).

13. Baxi and Dhanda, *Valiant Victims and Lethal Litigation,* p. 61.

14. Sheila Jasanoff, *Science at the Bar: Law, Science, and Technology in America* (Cambridge, MA: Harvard University Press, 1995), pp. 114–37.

15. See *Bano v. Union Carbide,* 273 F.3d 120 (2nd Cir. 2001).

16. Michael Wines, "Duke Energy to Pay Fine over Power Plant Violations," *New York Times,* September 10, 2015.

17. Andrew Lakoff, ed., *Disaster and the Politics of Intervention* (New York: SSRC/Columbia University Press, 2009).

18. Sheila Jasanoff, "Technologies of Humility: Citizen Participation in Governing Science," *Minerva* 41 (2003): 223–44.

Chapter 4: Remaking Nature

1. Paul J. Crutzen and Eugene F. Stoermer, "The 'Anthropocene,'" *Global Change Newsletter,* no. 41 (May 2000): 17–18, available at http://www.igbp.net/download/18.316f18321323470177580001401/1376383088452/NL41.pdf, accessed November 2015.

2. Charles E. Rosenberg, *No Other Gods: On Science and American Social Thought,* rev. ed. (Baltimore, MD: Johns Hopkins University Press, 1997), pp. 153–72.

3. Les Levidow and Susan Carr, *GM Food on Trial* (New York: Routledge, 2010).

4. Sheila Jasanoff, *Designs on Nature: Science and Democracy in Europe and the United States* (Princeton, NJ: Princeton University Press, 2005).

5. The International Agency for Research on Cancer has classified glyphosate as a probable cause of cancer. *IARC Monographs Volume 112: Evaluation of Five Organophosphate Insecticides and Herbicides,* March 20, 2015. For a dissenting voice, see Michael Specter, "Roundup and Risk Assessment," *New Yorker,* April 10, 2015.

6. U.S. Department of Agriculture, Economic Research Service, "Adop-

tion of Genetically Engineered Crops in the U.S.," http://www.ers.usda
.gov/data-products/adoption-of-genetically-engineered-crops-in-the-us
/recent-trends-in-ge-adoption.aspx#.UvcHMvb9rbk, accessed November
2014.

7. David A. Graham, "Rumsfeld's Knowns and Unknowns: The Intellectual History of a Quip," *Atlantic*, March 27, 2014.

8. David Barboza, "Gene-Altered Corn Changes Dynamics of Grain Industry," *New York Times*, December 11, 2000.

9. Colin A. Carter and Aaron Smith, "Estimating the Market Effect of a Food Scare: The Case of Genetically Modified Starlink Corn," *Review of Economics and Statistics* 89, no. 3 (2007): 522–33.

10. Amelia P. Nelson, "Legal Liability in the Wake of *StarLink*™: Who Pays in the End?," *Drake Journal of Agricultural Law* 7 (2002): 241–66.

11. Jasanoff, *Designs on Nature*, pp. 42–67.

12. Joel Tickner, ed., *Precaution, Environmental Science, and Preventive Public Policy* (Washington, DC: Island Press, 2003).

13. World Trade Organization, Agreement on Sanitary and Phytosanitary Measures, Article 5 (Assessment of Risk and Determination of the Appropriate Level of Sanitary or Phytosanitary Protection), sections 1, 2, and 7.

14. David Winickoff et al., "Adjudicating the GM Food Wars: Science, Risk, and Democracy in World Trade Law," *Yale Journal of International Law* 30 (2005): 81.

15. David A. Wirth, "The World Trade Organization Dispute over Genetically Modified Organisms: The Precautionary Principle Meets International Trade Law," *Vermont Law Review* 37, no. 4 (2013): 1187.

16. Growing research on gene flows indicated, however, that initial expectations about the degree of separation that needs to be maintained in order to prevent contamination between GM and non-GM crops were too optimistic.

17. Javier Lezaun, "Bees, Beekeepers and Bureaucrats: Parasitism and the Politics of Transgenic Life," *Environment and Planning D: Society and Space* 29 (2011): 738–58.

18. European Court of Justice (4th Chamber), *Monsanto SAS and Others v. Ministre de l'Agriculture et de la Pêche*, September 8, 2011.

19. Monsanto, "European Bans on MON810 Insect Protected GMO Corn Hybrids," http://www.monsanto.com/newsviews/pages/mon810
-background-information.aspx, accessed March 2014.

20. Kristina Hubbard, "Remember StarLink?," *Seed Broadcast Blog*, Febru-

ary 11, 2011, http://blog.seedalliance.org/2011/02/11/remember-starlink/, accessed March 2014.

21. Ellen Barry, "After Farmers Commit Suicide, Debts Fall on Families in India," *New York Times*, February 22, 2014.

22. J.L.P., "GMO Genocide?," *Economist*, March 13, 2014, http://www .economist.com/blogs/feastandfamine/2014/03/gm-crops-indian -farmers-and-suicide.

23. University of Cambridge, Research News, "New Evidence of Suicide Epidemic among India's 'Marginalised' Farmers," April 17, 2014, http://www.cam.ac.uk/research/news/new-evidence-of-suicide-epidemic -among-indias-marginalised-farmers.

Chapter 5: Tinkering with Humans

1. James D. Watson and Francis H. C. Crick, "A Structure for Deoxyribose Nucleic Acid," *Nature* 171 (1953): 737–38.

2. For example, the Council of Europe's 1997 legally binding Convention on Human Rights and Biomedicine prohibits germline genetic modification.

3. Sally Smith Hughes, "Making Dollars out of DNA: The First Major Patent in Biotechnology and the Commercialization of Molecular Biology, 1974–1980," *Isis* 92 (2001): 541–75.

4. Sheila Jasanoff, *Reframing Rights: Bioconstitutionalism in the Genetic Age* (Cambridge, MA: MIT Press, 2011).

5. Angelina Jolie, "My Medical Choices," *New York Times*, May 14, 2013.

6. Jennifer E. Reardon, *Race to the Finish: Identity and Governance in an Age of Genomics* (Princeton, NJ: Princeton University Press, 2004); Dorothy Roberts, *Fatal Invention: How Science, Politics, and Big Business Re-create Race in the Twenty-First Century* (New York: New Press, 2011); Jonathan Kahn, *Race in a Bottle: The Story of BiDil and Racialized Medicine in a Post-Genomic Age* (New York: Columbia University Press, 2013).

7. Genetic Information Nondiscrimination Act of 2008 (P.L. 110-233, 122 Stat. 881).

8. David Winickoff, "Genome and Nation: Iceland's Health Sector Database and Its Legacy," *Innovations* 1, no. 2 (Spring 2006): 80–105.

9. G. Owen Schafer, Ezekiel J. Emanuel, and Alan Wertheimer, "The Obligation to Participate in Biomedical Research," *Journal of the American Medical Association* 302, no. 1 (2009): 67–72.

10. National Bioethics Advisory Commission, *Cloning Human Beings* (Rockville, MD, 1997).

11. Sheila Jasanoff, J. Benjamin Hurlbut, and Krishanu Saha, "CRISPR Democracy: Gene Editing and the Need for Inclusive Deliberation," *Issues in Science and Technology*, Fall 2015, pp. 25–32.

12. Warnock Report.

13. Jodi Picoult, *My Sister's Keeper* (New York: Washington Square Press, 2004); Kazuo Ishiguro, *Never Let Me Go* (New York: Vintage, 2006).

14. Peter R. Brindsen, "Gestational Surrogacy," *Human Reproduction Update* 9, no. 5 (2003): 483–91.

15. European Court of Human Rights, Press Release, ECHR 185 (2014), June 26, 2014.

Chapter 6: Information's Wild Frontiers

1. *Riley v. California*, 573 U.S. ___ (2014).

2. In *Katz v. United States*, 389 U.S. 367 (1967), the Supreme Court held that a man suspected of placing illegal interstate gambling bets from a public telephone booth had a reasonable expectation that his conversations were private and would not be picked up by an electronic listening device attached to the outside of the phone booth. The case established that electronic eavesdropping was equivalent to a physical intrusion and that a phone booth was a protected space in the sense contemplated by the Fourth Amendment.

3. *Schmerber v. California*, 384 U.S. 757 (1966).

4. *Maryland v. King*, 569 U. S. ___ (2013).

5. As in many U.S. civil liberties cases, the four dissenting justices championed a radically different point of view. Surprisingly to some, Justice Antonin Scalia launched the most vigorous attack, excoriating the majority for having opened the door wide to abuse by the state: "Make no mistake about it: As an entirely predictable consequence of today's decision, your DNA can be taken and entered into a national DNA database if you are ever arrested, rightly or wrongly, and for whatever reason."

6. Sherry Turkle, "Always On: Always-On-You: The Tethered Self," in James E. Katz, ed., *Handbook of Mobile Communication Studies* (Cambridge, MA: MIT Press, 2008), pp. 121–39.

7. Cass Sunstein, "Shopping Made Psychic," *New York Times*, August 20, 2014.

8. *Griswold v. Connecticut*, 381 U.S. 479 (1965).

9. Alison Motluk, "Anonymous Sperm Donor Traced on Internet," *New Scientist*, November 3, 2005, http://www.newscientist.com/article/mg18 825244.200-anonymous-sperm-donor-traced-on-Internet.html, accessed August 2014.

10. Michal Kosinski, David Stillwell, and Thore Graepel, "Private Traits and Attributes Are Predictable from Digital Records of Human Behavior," *Proceedings of the National Academy of Sciences*, March 11, 2013, http://www.pnas.org/content/early/2013/03/06/1218772110.full.pdf.

11. Nicole Perlroth, "Hackers Say They Have Released Ashley Madison Files," *New York Times*, August 19, 2015.

12. Maria Aspan, "How Sticky Is Membership on Facebook? Just Try Breaking Free," *New York Times*, February 11, 2008.

13. Robinson Meyer, "Everything We Know about Facebook's Secret Mood Manipulation Experiment," *Atlantic*, June 28, 2014.

14. Adam D. I. Kramer, Jamie E. Guillory, and Jeffrey T. Hancock, "Experimental Evidence of Massive-Scale Emotional Contagion through Social Networks," *Proceedings of the National Academy of Sciences* 111, no. 24 (June 17, 2014). From the abstract: "When positive expressions were reduced, people produced fewer positive posts and more negative posts; when negative expressions were reduced, the opposite pattern occurred."

15. Adam D. I. Kramer, June 29, 2014, https://www.facebook.com/akramer/posts/10152987150867796.

16. Kashmir Hill, "After the Freak-Out over Facebook's Emotion Manipulation Study, What Happens Now?," *Forbes*, July 10, 2014, http://www.forbes.com/sites/kashmirhill/2014/07/10/after-the-freak-out-over-face books-emotion-manipulation-study-what-happens-now/.

17. Viktor Mayer-Schönberger, *Delete: The Virtue of Forgetting in the Digital Age* (Princeton, NJ: Princeton University Press, 2011).

18. *New York Times v. United States*, 403 U.S. 713 (1971).

19. Dan Bilefsky, "Indignation in Europe over Claims That U.S. Spied on Merkel's Phone," *New York Times*, October 24, 2013; Mark Landler, "Merkel Signals That Tension Persists over U.S. Spying," ibid., May 2, 2014.

20. Sheila Jasanoff, *Science and Public Reason* (Abingdon, Oxon: Routledge, 2012).

21. M. G. Zimeta, "Don't Be Evil: Google, Alphabet, and Machiavelli," *Paris Review*, August 12, 2015, http://www.theparisreview.org/blog/2015/08/12/dont-be-evil/, accessed January 2016.

22. Matt Rosoff, "Is Google a Monopoly? 'We're in That Area,' Admits Schmidt," *Business Insider*, September 21, 2011, http://www.businessinsider

.com/is-google-a-monopoly-were-in-that-area-admits-schmidt-2011-9#ixzz3BEHV6le9, accessed August 2014.

23. James Kanter, "Google's European Antitrust Woes Are Far from Over," *New York Times,* June 22, 2014.

24. Benedict Anderson, *Imagined Communities: Reflections on the Origin and Spread of Nationalism* (London: Verso, 1983).

25. *In Re High-Tech Employee Antitrust Litigation,* 11-CV-0250, 2014 U.S. Dist. LEXIS 110064 (N.D. Cal. Aug. 8, 2014).

26. David Streitfeld, "Court Rejects Deal on Hiring in Silicon Valley," *New York Times,* August 8, 2014.

27. David Streitfeld and Maria Wollan, "Tech Rides Are Focus of Hostility in Bay Area," *New York Times,* January 31, 2014.

28. Glenn Greenwald and Ewan MacAskill, "NSA Prism Program Taps in to User Data of Apple, Google and Others," *Guardian,* June 6, 2013.

29. Craig Timberg, "U.S. Threatened Massive Fine to Force Yahoo to Release Data," *Washington Post,* September 11, 2014.

30. "Breaking Down Apple's iPhone Fight With the U.S. Government," *New York Times,* March 21, 2016, http://www.nytimes.com/interactive/2016/03/03/technology/apple-iphone-fbi-fight-explained.html, accessed April 2016.

31. Tim Cook, "A Message to Our Customers," February 16, 2016, http://www.apple.com/customer-letter/, accessed April 2016.

32. Landler, "Merkel Signals That Tension Persists over U.S. Spying."

33. *Google Spain SL, Google Inc. v. Agencia Española de Protección de Datos (AEPD), Mario Costeja González,* Case C-131/12, May 13, 2014, press release no, 70/14, at http://curia.europa.eu/jcms/upload/docs/application/pdf/2014-05/cp140070en.pdf, accessed August 2014.

34. David Streitfeld, "European Court Lets Users Erase Records on Web," *New York Times,* May 13, 2014.

Chapter 7: Whose Knowledge, Whose Property?

1. Kathy L. Hudson and Francis S. Collins, "Biospecimen Policy: Family Matters," *Nature* 500 (August 8, 2013): 141–42.

2. Rebecca Skloot, *The Immortal Life of Henrietta Lacks* (New York: Crown, 2010).

3. Arthur Caplan, "NIH Finally Makes Good with Henrietta Lacks' Family—And It's About Time, Ethicist Says," NBC News, August 7, 2013,

http://www.nbcnews.com/health/health-news/nih-finally-makes-good-henrietta-lacks-family-its-about-time-f6C10867941, accessed August 2014.

4. Ewen Callaway, "Deal Done over HeLa Cell Line," *Nature* 500 (August 7, 2013): 132–33, http://www.nature.com/news/deal-done-over-hela-cell-line-1.13511, accessed August 2014.

5. Quoted in Rebecca Skloot, "The Immortal Life of Henrietta Lacks, the Sequel," *New York Times*, March 23, 2013.

6. "Thomas Jefferson to Isaac McPherson, 13 August 1813," Founders Online, National Archives, http://founders.archives.gov/documents/Jefferson/03-06-02-0322, accessed August 2014.

7. The 1790 act provided a slightly longer list: "any useful art, manufacture, engine, machine, or device, or any improvement therein not before known or used."

8. Edward C. Walterscheid, "Patents and the Jeffersonian Mythology," *John Marshall Law Review* 29, no. 1 (1995): 293.

9. 35 U.S. Code § 101 – Inventions patentable.

10. Callaway, "Deal Done over HeLa Cell Line."

11. Rebecca Skloot, "Taking the Least of You," *New York Times*, April 16, 2006.

12. *Moore v. Regents of the University of California*, 51 Cal. 3d 120 (1990).

13. "Who Owns Your Body? The Catalona Case: Patients Lose Lawsuit to Claim Their Tissues," http://www.whoownsyourbody.org/catalona.html.

14. *Washington University v. Catalona*, 490 F.3d 667 (8th Cir. 2007).

15. Lisa C. Edwards, "Tissue Tug-of-War: A Comparison of International and U.S. Perspectives on the Regulation of Human Tissue Banks," *Vanderbilt Journal of Transnational Law* 41 (2008): 639–75.

16. Daniel J. Kevles, "Ananda Chakrabarty Wins a Patent: Biotechnology, Law, and Society," *Historical Studies in the Physical and Biological Sciences* 25, pt. 1 (1994): 111–35.

17. *Diamond v. Chakrabarty*, 447 U.S. 303 (1980).

18. Brief on Behalf of Genentech, Inc., *Amicus Curiae*, Supreme Court of the United States, October Term, 1979, No. 79-136, p. 16. See also Sheila Jasanoff, "Taking Life: Private Rights in Public Nature," in Kaushik Sunder Rajan, ed., *Lively Capital: Biotechnologies, Ethics, and Governance in Global Markets* (Durham, NC: Duke University Press, 2012), pp. 155–83.

19. Sheila Jasanoff, *Science at the Bar: Law, Science, and Technology in America* (Cambridge, MA: Harvard University Press, 1995), pp. 36–39.

20. Brief on Behalf of the Peoples Business Commission, Amicus Curiae,

Supreme Court of the United States, October Term, 1979, No. 79-136, p. 5.

21. Edmund L. Andrews, "U.S. Resumes Granting Patents on Genetically Altered Animals," *New York Times*, February 3, 1993.

22. Directive 98/44/EC of the European Parliament and of the Council of July 6, 1998, on the legal protection of biotechnological inventions. Article 6.1 provides, "Inventions shall be considered unpatentable where their commercial exploitation would be contrary to ordre public or morality."

23. *President and Fellows of Harvard College v. Canada (Commissioner of Patents)* 2002 SCC 76, para. 163.

24. William Safire, "Language: Centaurs, Chimeras, and Humanzees," *New York Times*, May 23, 2005.

25. Stephen R. Munzer, "Human-Nonhuman Chimeras in Embryonic Stem Cell Research," *Harvard Journal of Law and Technology* 21 (2007): 123.

26. *Parke-Davis & Co. v. H. K. Mulford Co.*, 189 F. 95 (2011).

27. Heidi Ledford, "Tania Simoncelli: Gene Patent Foe," in "365 Days: *Nature's* 10," *Nature*, December 18, 2013, http://www.nature.comnews/365-days -nature-s-10-1.14367#/Tania.

28. *Association of Molecular Pathology v. U.S. Patent & Trademark Office*, 702 F. Supp. 2d 181 (S.D.N.Y. 2010), at p. 184.

29. *Association of Molecular Pathology v. U.S. Patent & Trademark Office*, 689 F. 3d 1303, 1326 (CAFC 2012).

30. Brief for *Amicus Curiae* Eric S. Lander in Support of Neither Party, *Association of Molecular Pathology v. U.S. Patent & Trademark Office*, No. 12-398 (2013).

31. Paul Rabinow, *Making PCR: A Story of Biotechnology* (Chicago: University of Chicago Press, 1996).

32. Cory Hayden, *When Nature Goes Public: The Making and Unmaking of Bioprospecting in Mexico* (Princeton, NJ: Princeton University Press, 2003).

33. Bernd Siebenhüner and Jessica Supplie, "Implementing the Access and Benefit-Sharing Provisions of the CBD: A Case for Institutional Learning," *Ecological Economics* 53 (2005): 507–22.

34. According to section 3(d), "The mere discovery of a new form of a known substance which does not result in the enhancement of the known efficacy of that substance" does not entitle the manufacturer to a patent.

35. *Novartis v. Union of India*, Supreme Court of India, April 1, 2013.

36. Ibid., para. 36.

37. Gardiner Harris, "Maker of Costly Hepatitis C Drug Sovaldi Strikes Deal on Generics for Poor Countries," *New York Times*, September 14, 2014.

Chapter 8: Reclaiming the Future

1. Ferris Jabr, "The Science Is In: Elephants Are Even Smarter Than We Realized," *Scientific American*, February 26, 2014.

2. Desmond Morris, "Can Jumbo Elephants Really Paint?," *Mail Online*, February 21, 2009, http://www.dailymail.co.uk/sciencetech/article-1151283 /Can-jumbo-elephants-really-paint--Intrigued-stories-naturalist-Des mond-Morris-set-truth.html, accessed September 2014.

3. Office of Technology Assessment Act, Public Law 92-484, 92nd Congress, H.R. 10243, October 13, 1972.

4. Bruce Bimber, *The Politics of Expertise in Congress: The Rise and Fall of the Office of Technology Assessment* (Albany: State University of New York Press, 1996).

5. Colin Norman, "O.T.A. Caught in Partisan Crossfire," *Technology Review*, October/November 1977.

6. Bimber, *The Politics of Expertise in Congress*, p. 77.

7. American Academy of Arts and Sciences, *Restoring the Foundation: The Vital Role of Research in Preserving the American Dream* (Cambridge, MA: AAAS, 2014), p. 65.

8. Tom Wicker, "In the Nation: Two Spacey Schemes," *New York Times*, May 11, 1984.

9. Frank von Hippel, "Attacks on Star Wars Critics a Diversion," *Bulletin of the Atomic Scientists*, April 1985, pp. 8–10.

10. For more details, see Sheila Jasanoff, *Designs on Nature: Science and Democracy in Europe and the United States* (Princeton, NJ: Princeton University Press, 2005).

11. Office of Technology Assessment, *New Developments in Biotechnology: Field-Testing Engineered Organisms: Genetic and Ecological Issues* (Washington, DC: Government Printing Office, 1988), p. 20.

12. Office of Technology Assessment, *New Developments in Biotechnology*, p. 26.

13. Jim Robbins, "The Year the Monarch Didn't Appear," *New York Times*, November 22, 2013.

14. Kenneth A. Worthy, Richard C. Strohman, Paul R. Billings, et al., "Agricultural Biotechnology Science Compromised: The Case of Quist and Chapela," in Daniel L. Kleinman, Abby J. Kinchy, and Jo Handelsman, eds., *Controversies in Science and Technology: From Maize to Menopause* (Madison: University of Wisconsin Press, 2005), pp. 135–49.

15. Johan Schot and Arie Rip, "The Past and Future of Constructive Technology Assessment," *Technological Forecasting and Social Change* 54 (1996): 257.

16. The Nuremberg Code, http://www.hhs.gov/ohrp/archive/nurcode.html, accessed October 2014.

17. Leon R. Kass, "The Wisdom of Repugnance," *New Republic,* June 2, 1997, pp. 17–26.

18. See, e.g., Presidential Commission for the Study of Bioethical Issues, *New Directions: The Ethics of Synthetic Biology and Emerging Technologies* (Washington, DC: 2010).

19. National Human Genome Research Institute, "ELSI Planning and Evaluation History," http://www.genome.gov/10001754, accessed October 2014.

20. Sheila Jasanoff, "Constitutional Moments in Governing Science and Technology," *Science and Engineering Ethics* 17, no. 4 (2011): 621–38.

21. Jennifer Gollan, "Lab Fight Raises U.S. Security Issues," *New York Times,* October 22, 2011.

22. Ibid.

23. Deutscher Ethikrat, http://www.ethikrat.org/about-us/ethics-council-act, accessed October 2014.

24. German Ethics Council, *Human-Animal Mixtures in Research* (2011), p. 98.

25. See complete text at http://nuffieldbioethics.org/about/#sthash.Ex2ugB61 .dpuf.

26. John H. Evans, *The History and Future of Bioethics: A Sociological View* (New York: Oxford University Press, 2012).

27. Sheila Jasanoff and Sang-Hyun Kim, eds., *Dreamscapes of Modernity: Sociotechnical Imaginaries and the Fabrication of Power* (Chicago: University of Chicago Press, 2015).

28. Jason Chilvers and Matthew Kearnes, eds., *Remaking Participation: Science, Environment and Emergent Publics* (Abingdon, Oxon: Routledge, 2015).

29. Proceedings from the Congressional Record of March 12, 1946 (Administrative Procedure), http://www.justice.gov/sites/default/files/jmd/legacy /2013/11/19/proceedings-05-1946.pdf, accessed October 2014.

30. Sheila Jasanoff, *Science and Public Reason* (Abingdon, Oxon: Routledge-Earthscan, 2012).

31. Jasanoff, "Constitutional Moments."

32. Robert Doubleday and Brian Wynne, "Despotism and Democracy in the United Kingdom: Experiments in Reframing Citizenship," in Sheila Jasanoff, ed., *Reframing Rights: Bioconstitutionalism in the Genetic Age* (Cambridge, MA: MIT Press, 2011), pp. 239–62.

33. The Agricultural and Environment Biotechnology Commission (AEBC) was a twenty-member body established by Tony Blair's Labour govern-

ment in response to the controversies of the late 1990s. AEBC's role was to advise the UK government on the ethical and social as well as the scientific dimensions of GM crops. It was disbanded in 2005 when its views increasingly diverged from the government's. See details at http://webarchive.nationalarchives.gov.uk/20100419143351/http://www.aebc.gov.uk/aebc/index.shtml, accessed November 2014.

34. Bernard Dixon, "Not Yet a GM Nation," *Current Biology* 13, no. 21 (2003): R819–R820, http://www.cell.com/current-biology/pdf/S0960-9822%2803%2900760-7.pdf, accessed November 2014.

35. Department for Environment, Food and Rural Affairs, "The GM Public Debate: Lessons Learned from the Process," March 2004, http://webarchive.nationalarchives.gov.uk/20081023141438/http://www.defra.gov.uk/environment/gm/crops/debate/pdf/gmdebate-lessons.pdf, accessed November 2014.

36. Matthew Kearnes, Robin Grove-White, Phil MacNaghten, James Wilsdon, and Brian Wynne, "From Bio to Nano: Learning Lessons from the UK Agricultural Biotechnology Controversy," *Science as Culture* 15, no. 4 (2006): 301–2.

37. David Noble, *America by Design: Science, Technology, and the Rise of Corporate Capitalism* (New York: Alfred A. Knopf, 1977).

Chapter 9: Invention for the People

1. "Human Cloning at Last," Runners Up, Breakthrough of the Year, *Science* 342 (December 20, 2013).

2. National Bioethics Advisory Commission, *Cloning Human Beings: Report and Recommendations of the National Bioethics Advisory Commission* (Rockville, MD: NBAC, 1997); and Presidential Commission for the Study of Bioethical Issues, *New Directions: The Ethics of Synthetic Biology and Emerging Technologies* (Washington, DC: 2010).

3. David Guston, "Understanding 'Anticipatory Governance,'" *Social Studies of Science* 44, no. 2 (2013): 218–42.

4. Andrew Pollack, "U.S.D.A. Approves Modified Potato. Next Up: French Fry Fans," *New York Times*, November 7, 2014.

5. D. Rowan, "On the Exponential Curve: Inside Singularity University," *Wired*, May 6, 2013, http://www.wired.co.uk/magazine/archive/2013/05/singularity-university/on-the-exponential-curve, accessed November 2014.

6. Elta Smith, "Corporate Imaginaries of Biotechnology and Global Governance: Syngenta, Golden Rice and Corporate Social Responsibility," in Sheila Jasanoff and Sang-Hyun Kim, eds., *Dreamscapes of Modernity: Sociotechnical Imaginaries and the Fabrication of Power* (Chicago: University of Chicago Press, 2015), pp. 254–76.

7. Daryl Collins, Jonathan Morduch, Stuart Rutherford, and Orlanda Ruthven, *Portfolios of the Poor: How the World's Poor Live on $2 a Day* (Princeton, NJ: Princeton University Press, 2009).

8. Executive Office of the President, President's Council of Advisors on Science and Technology, *Big Data and Privacy: A Technological Perspective* (Washington, DC: March 2014).

9. *Novartis v. Union of India*, Supreme Court of India, April 1, 2013, para. 38.

10. Jon Ronson, *So You've Been Publicly Shamed* (New York: Riverhead, 2015).

INDEX

Page numbers followed by *n* refer to material in notes.

Index